AF743861

Environmental Remediation Technologies, Regulations and Safety

nova science publishers

www.novapublishers.com

Copyright © 2022 by Nova Science Publishers, Inc.
DOI: https://doi.org/10.52305/AOXA6854

All rights reserved. No part of this book may be reproduced, stored in a retrieval system or transmitted in any form or by any means: electronic, electrostatic, magnetic, tape, mechanical photocopying, recording or otherwise without the written permission of the Publisher.

We have partnered with Copyright Clearance Center to make it easy for you to obtain permissions to reuse content from this publication. Simply navigate to this publication's page on Nova's website and locate the "Get Permission" button below the title description. This button is linked directly to the title's permission page on copyright.com. Alternatively, you can visit copyright.com and search by title, ISBN, or ISSN.

For further questions about using the service on copyright.com, please contact:
Copyright Clearance Center
Phone: +1-(978) 750-8400 Fax: +1-(978) 750-4470 E-mail: info@copyright.com.

NOTICE TO THE READER

The Publisher has taken reasonable care in the preparation of this book, but makes no expressed or implied warranty of any kind and assumes no responsibility for any errors or omissions. No liability is assumed for incidental or consequential damages in connection with or arising out of information contained in this book. The Publisher shall not be liable for any special, consequential, or exemplary damages resulting, in whole or in part, from the readers' use of, or reliance upon, this material. Any parts of this book based on government reports are so indicated and copyright is claimed for those parts to the extent applicable to compilations of such works.

Independent verification should be sought for any data, advice or recommendations contained in this book. In addition, no responsibility is assumed by the Publisher for any injury and/or damage to persons or property arising from any methods, products, instructions, ideas or otherwise contained in this publication.

This publication is designed to provide accurate and authoritative information with regard to the subject matter covered herein. It is sold with the clear understanding that the Publisher is not engaged in rendering legal or any other professional services. If legal or any other expert assistance is required, the services of a competent person should be sought. FROM A DECLARATION OF PARTICIPANTS JOINTLY ADOPTED BY A COMMITTEE OF THE AMERICAN BAR ASSOCIATION AND A COMMITTEE OF PUBLISHERS.

Additional color graphics may be available in the e-book version of this book.

Library of Congress Cataloging-in-Publication Data

ISBN: 978-1-68507-974-1

Published by Nova Science Publishers, Inc. † New York

Contents

Preface vii

Chapter 1 **Mathematical Modeling of Wind Turbines** 1
Mustafa Şahin and İlkay Yavrucuk

Chapter 2 **Wind Turbine Controller Design** 35
Mustafa Şahin and İlkay Yavrucuk

Chapter 3 **Energy Potential Wind Facility Calculation from Field Measurements and Technical Parameters of the Wind Farm Using Weibull Distribution Methodology** 67
Jesús Ordoñez-Guerrero

Chapter 4 **State Feedback Controller Design for Batteries Participating in Load Frequency Control in Power Networks with Wind Generation** 103
Semaria Ruiz and Daniel Icaza

Chapter 5 **Wind Turbines for Energy Production in Organic Buildings** 147
Daniel Icaza, David Borge-Diez, Semaria Ruiz and Santiago Pulla-Galindo

Chapter 6 **Wind Distributed Generation and Desalination: An Application to a Rural Microgrid** 181
David Borge-Diez, Francisco José García-Moya and Enrique Rosales-Asensio

Chapter 7 **Implementation Analysis of the Largest Wind Project in Ecuador** 199
Santiago Pulla Galindo, David Borge-Diez and Daniel Icaza

About the Editors 231

List of Contributors 233

Index 235

Preface

This book constitutes a reference book for innovative applications of wind energy including modelling of these wind systems and innovative applications for distributed generation and grid-connected systems. The book includes a description of wind systems modelling methods and control systems and explores innovative applications for distributed generation systems, water desalination and large-scale plants in regions with specific characteristics, such as Ecuador.

This book constitutes a reference for researchers and technicians in the wind energy field looking for reference knowledge in wind modelling but also for practical and novel applications.

Editors
David Borge-Diez
Department of Electrical, Systems and Automation Engineering,
University of León, León, Spain

Daniel Icaza
Catholic University of Cuenca, Ecuador. Center for Research, Innovation and Technology Transfer CIITT, Lighting Technology Research Laboratory CIITT. GIRVyP Group

Chapter 1

Mathematical Modeling of Wind Turbines

Mustafa Şahin[1,*], PhD and İlkay Yavrucuk[2], PhD

[1]METU Center for Wind Energy Research, Middle East Technical University, Ankara, Turkey

[2]Department of Aerospace Engineering, Middle East Technical University, Ankara, Turkey

Abstract

This chapter analyzes the mathematical modeling of a rigidly structured Horizontal Axis Wind Turbine (HAWT) System based on Blade Element Momentum (BEM) theory. The turbine model includes several aerodynamic corrections such as turbulent wake state, skewed wake, hub and tip losses, etc. It has the abilities of collective and individual blade pitching, nacelle yawing, rotor precone and tilt angle definitions. The MS (Mustafa Şahin) Bladed Wind Turbine Simulation Model, which utilizes the same theoretical approach and properties in this chapter, is given as an example turbine model with its implementation, operation principle, and validation using NREL experimental turbines such Phase II and Phase III.

Keywords: wind turbine mathematical modeling, Blade Element Momentum (BEM) theory, skewed wake, turbulent wake, hub and tip losses, turbine coordinate systems, MS Bladed Model, implementation, operation principle, and validation

[*]Corresponding Author's Email: musahin33@gmail.com; musahin@metu.edu.tr.

In: Wind Turbines
Editors: David Borge-Diez and Daniel Icaza
ISBN: 978-1-68507-974-1

© 2022 Nova Science Publishers, Inc.

Introduction

Wind turbines are one of the prominent systems using clean energy for electricity generation. They may be designed and manufactured in different sizes and appearances (Figure 1). The most commonly employed turbine type in wind farms is the Horizontal Axis Wind Turbine (HAWT) due to its efficiency (Pao and Johnson 2009). The number of these turbines increases every year around the world. Now, more than 90 countries use turbines to supply their electrical demand, and the worldwide installed cumulative capacity has exceeded 650 GW (GWEC 2019). While onshore turbines are commonly seen in wind farms, offshore turbines are gaining popularity in the last decade, particularly for the last two years. Nowadays, the share of offshore turbines is around 5% of the total installed capacity in the world (GWEC 2021). China and the USA are the leading countries and tend to increase their cumulative capacity annually. More countries such as India, Spain, France, Brazil, United Kingdom, Turkey, etc. employ wind turbines.

A typical HAWT system includes a rotor (a hub with blades), a drive train (low and high-speed shafts, a gearbox, couplings, a mechanical brake, and an electrical generator), a nacelle, and a mainframe (housing, bedplate, and yaw system), electrical and control systems, a foundation and a tower that keeps the whole system up in the airflow (Singh and Santoso 2011). Of all the components, the turbine rotor is considered to be the dominating part of the turbine characteristics. Turbines with efficient blades can extract high energy from wind, which is eventually turned into a high amount of electrical power (Sahin and Yavrucuk 2017).

An area for wind turbine research and development is modeling. A wind turbine model allows designers to predict turbine performance, simulate turbines, and see their behavior under various conditions, such as turbulent winds at various levels, different air densities, structural angles, etc. The turbine behaviour can be observed as a function of time in terms of various turbine parameters. Using the turbine model, turbine controllers, protection algorithms, and so on can be developed. For instance, in the next chapter, the design and simulation of baseline controllers are carried out using the MS Bladed Model for the NREL 5 MW turbine. Also, new ideas, methodologies, control algorithms can be designed, implemented, and tested on the model. As an example, a wind turbine protection algorithm is developed and simulated in Ref (Sahin and Yavrucuk 2020) using the MS Bladed Model.

A common method to develop such turbine models is through the use of Blade Element Momentum (BEM) theory. This theory is often utilized due to its ease of implementation and computational effectiveness. In turbine modeling, along with BEM theory, some aerodynamic corrections such as tip and hub loss corrections, turbine wake state operation correction, skewed wake correction, etc. are required to represent a more accurate turbine operation in the field. Wind, turbine, and rotor geometric properties as well as blade airfoil data are introduced to the BEM-based turbine simulation model as inputs. Accordingly, the turbine model generates outputs related to the employed turbine, such as generated power, thrust, tip speed ratio, or TSR, power coefficient, C_p and so on.

In this chapter, the modeling aspects of HAWT systems based on BEM theory are investigated. Firstly, HAWT modeling is examined. The HAWT model here has the properties of a blade pitching, nacelle yawing, precone and tilt angle definitions, etc. The same approach is used by the corresponding author to develop the MS (Mustafa Sahin) Bladed Simulation Model in the literature (Sahin 2018; Sahin and Yavrucuk 2019, 2020). For this reason, in this chapter, implementation, operation principle, and validation of the MS Bladed Model are given as an example.

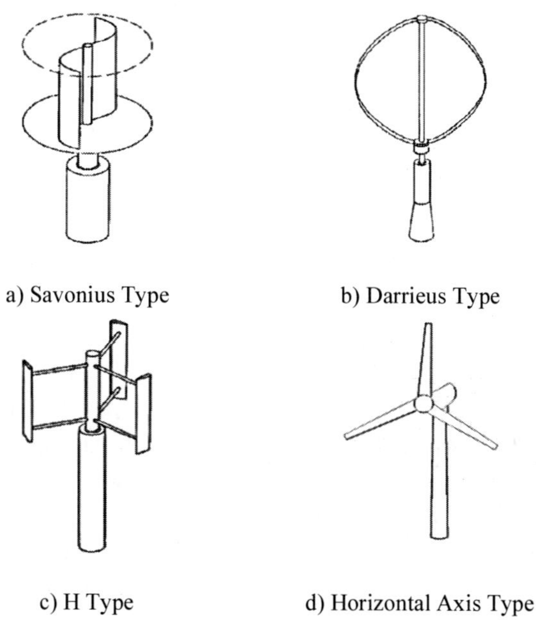

Figure 1. Wind turbine types (Erich 2006).

The rest of this chapter is organized as follows: Next section defines HAWT modeling, which is based on BEM theory and uses particular coordinate systems, and aerodynamic correction models. Afterwards, the implementation of the MS Bladed Model is briefly explained using MATLAB and Simulink programs. The model operation principle is defined. Model validation is investigated in detail using NREL experimental turbines. The last section is left for the conclusion part.

Wind Turbine Modeling

In this section, the modeling of a Horizontal Axis Wind Turbine system is discussed. The modeling can be implemented using computer programs such as MATLAB and Simulink. The main structure of the model is BEM theory. However, it includes several aerodynamic corrections such as hub and tip losses, turbulent wake state, and skewed wake rotation. It also employs particular coordinate systems to simulate the behavior of turbines with precone angle, nacelle/rotor tilt angle, etc. Here, the aerodynamic calculations of the turbine model are similar to those of PROP code(Wilson and Lissaman 1974), PROPID (M. S. Selig 2012), Wt_Perf (Platt and Buhl 2012), AeroDyn (Moriarty and Hansen 2005) programs. The HAWT model considers the turbine consisting of rigid structures such as blades, shaft(s), tower, etc., and employs an aerodynamic turbine rotor, a simple variable torque generator model, and a gearbox between these components. The following are the details.

For wind turbine modeling or aerodynamic calculations, BEM theory is a widely used method. This theory is a combination of blade element and momentum theories (Manwell, McGowan, and Rogers 2009). In blade element theory, the turbine blades are divided into the desired number of elements or sections (Figure 2a), and then aerodynamic lift and drag forces at each blade element are obtained as a function of blade geometry. Later, these aerodynamic forces are used to acquire the thrust force and torque of the turbine rotor. Taking into account the turbine rotor speed, turbine power can also be obtained. In momentum theory, however, an annular control volume approach is used and therefore wind turbine is considered as a rotating actuator disk in a stream tube (Figure 2b). The turbine thrust force is calculated from the linear momentum conservation principle, while the turbine rotor torque is obtained from the angular momentum theory. For the Blade Element Momentum (BEM) theory to be valid, there are some important restrictions or assumptions (Manwell, McGowan, and Rogers 2009). For example, the direction of the freestream wind flow is required to be perpendicular to the rotor disc swept by the blades. The wind flow is required to pass through the turbine blades, smoothly (Moriarty and Hansen 2005).

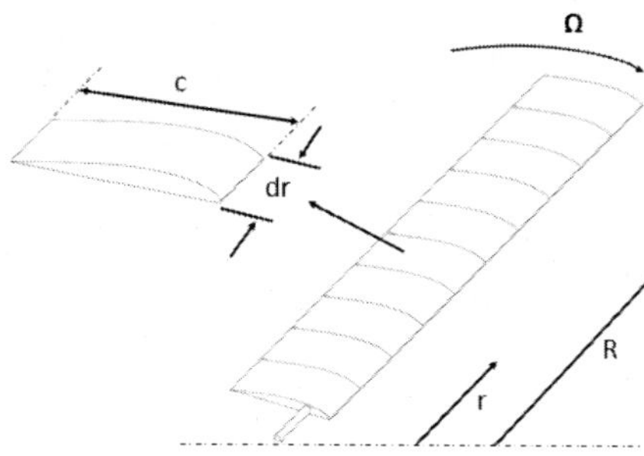

a)

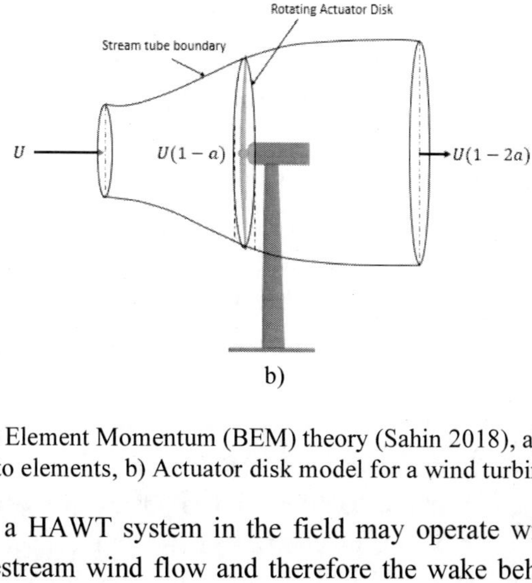

b)

Figure 2. Blade Element Momentum (BEM) theory (Sahin 2018), a) Division of a turbine blade into elements, b) Actuator disk model for a wind turbine.

However, a HAWT system in the field may operate with a yaw angle towards a freestream wind flow and therefore the wake behind the rotor is skewed with an angle, referred to as the skewed wake phenomenon (Moriarty and Hansen 2005). In such an operation, the performance of the turbine rotor is reduced and the prediction of BEM theory is to be corrected since the first assumption above is violated. Likewise, the same phenomenon occurs when the turbine rotor is given a rotor/nacelle tilt angle. Also, the turbine rotor may operate in the turbulent wake state, i.e., with the high axial induction factor, in which case, the airflow cannot pass the rotor blades, smoothly, and the second assumption above is broken. Accordingly, the turbine operates in the turbulent wake state at a high Tip Speed Ratio or TSR. Here, TSR is defined as the ratio of blade tip speed to the freestream wind speed (Burton et al. 2001). When turbine TSR is getting higher, the permeability of the turbine rotor to the airflow is reduced and a turbulent flow behind the rotor becomes dominant. This results in a larger thrust coefficient than the one predicted by the BEM theory, which also requires an aerodynamic correction. Furthermore, the BEM theory does not include the adverse effects of root and blade tip vortices in its calculations.

Additionally, to simulate a more modern turbine with blade pitching capability, collectively or individually, nacelle yawing capability, with the properties of defining the nacelle tilt and precone angles, with the changes of each blade attitude in time, i.e., rotor rotation, particular coordinate systems are also to be introduced to the HAWT model. Figure 3 shows coordinate systems that are utilized here for the HAWT modeling. The inertial and wind-aligned coordinate systems (Figure 3a) have origins at the center of the turbine tower base. In the inertial coordinate system, the x_i-axis may point to any direction when used consistently, but the convenient direction should be the main wind direction. The wind angle, β_{wind}, is positive when there is a rotation around +z axis. In the wind-aligned coordinate systems, the x_w-axis shows the main wind direction. Both z_i and z_w axes are along the tower. The origin of the yaw-aligned coordinate system (Figure 3b) is at the center of the tower top, i.e., at the yaw bearing, and rotates about the tower. The x_y-axis lies along the projection of the turbine rotor shaft in the horizontal plane or is aligned with the rotor shaft when there is no nacelle/rotor tilt angle. The angle between the x_y and x_w axes is referred to as the turbine yaw angle, Ψ. The wind and yaw-aligned coordinate systems (Figure 3b) share the same z-axis direction. Rotation about the +z direction is a positive yaw angle. However, there is a certain height, i.e., tower length, between the origins of these two coordinate systems. The y_y-axis follows from the right-hand rule, whereas z_y-axis is along the tower. The hub-aligned coordinate system (Figure 3c) has an origin located at the hub center. It is fixed to the hub center and does not rotate with the turbine rotor, but rotates with a yaw angle. The angle between the x_y and x_h axes is referred to as nacelle/rotor tilt angle, θ. To tilt the rotor up for an upwind turbine corresponds to a positive tilt angle +y axis. In the hub-aligned coordinate system, the x_h-axis lies along the rotor shaft toward the nominal downwind direction, or it is aligned with x_y when there is no nacelle tilt angle. The yaw and hub-aligned coordinate systems share a common y-axis. The z_h-axis, on the other hand, follows from the right-hand rule (Ning 2013; Sahin 2018).

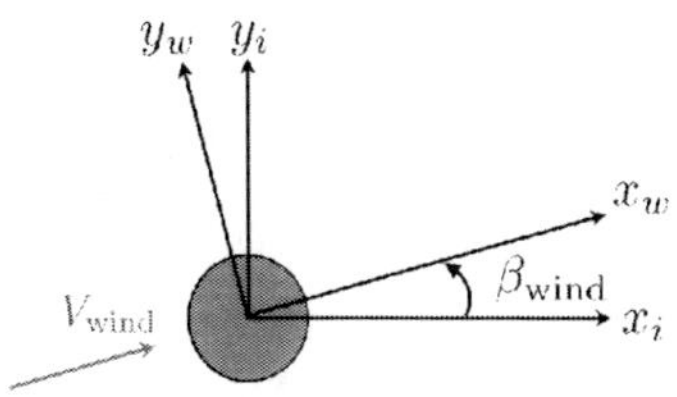

a) Inertial and wind-aligned coordinate systems

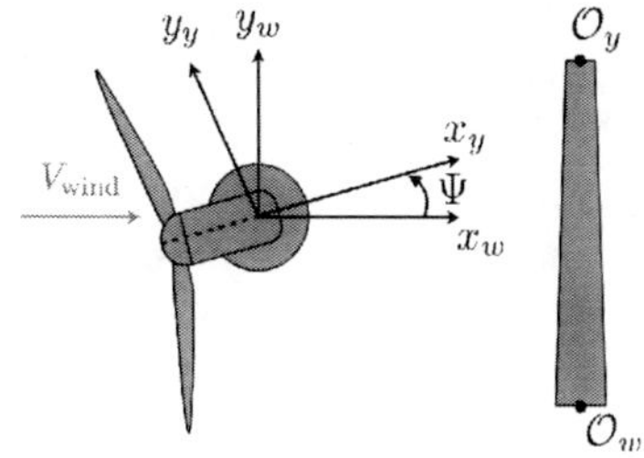

b) Wind-aligned and yaw-aligned coordinate systems

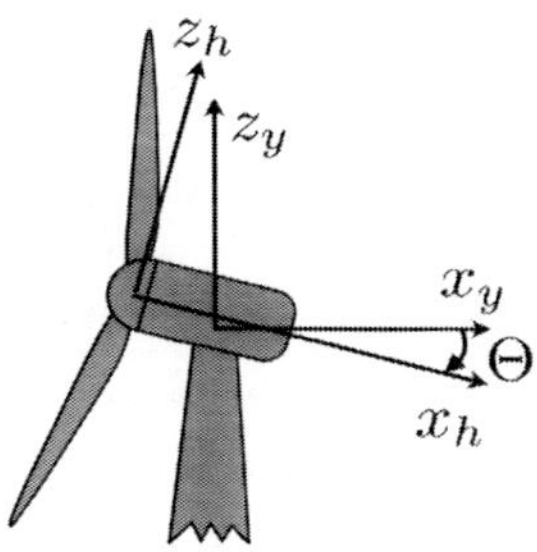

c) Yaw-aligned and hub-aligned coordinate systems

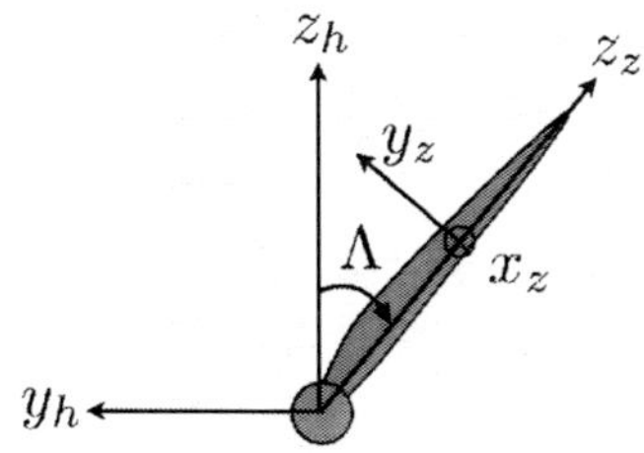

d) Hub-aligned and azimuth-aligned coordinate systems

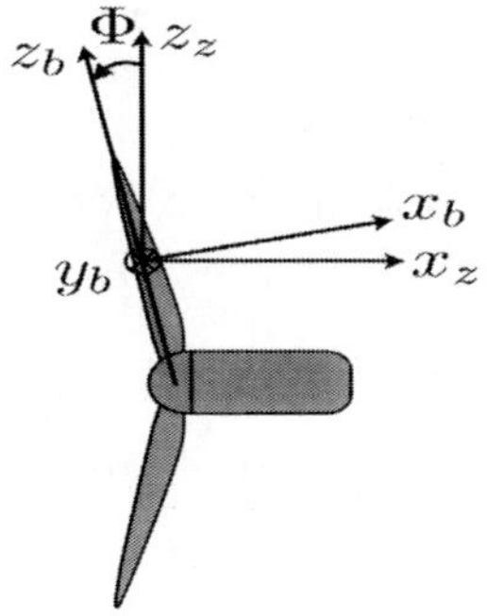

e) Azimuth-aligned and blade-aligned coordinate systems

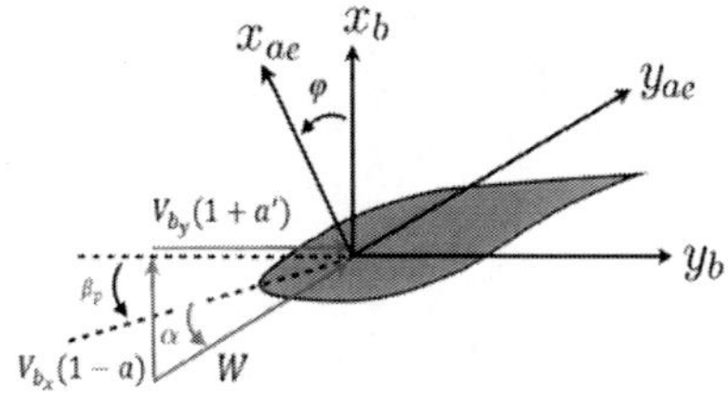

f) Blade-aligned and aerodynamic-aligned coordinate systems

Figure 3. HAWT coordinate systems (Ning 2013; Sahin 2018).

The azimuth and blade-aligned coordinate systems (Figure 3e) are locally fixed to each blade element and rotate with the angular movement of the rotor. The azimuth angle, Λ, is positive when there is a rotation about the +x-axis. The precone angle, Φ is negative when there is counterclockwise rotation about +y direction pointing into the page. This negative precone angle keeps the blades away from the tower for an upwind-oriented HAWT system. Lastly, the aerodynamic coordinate system (Figure 3f) is a locally defined coordinate system at the blade pitch axis, i.e., quarter cord of each airfoil. It is the coordinate system where the aerodynamic lift and drag forces are defined at every blade element. The coordinate system rotates with the turbine rotor blades. The x_{ae}-axis follows from the right-hand-rule. The y_{ae}-axis lies along the effective wind speed. The z_{ae}-axis points into the page in increasing radius and lies along the blade pitch axis, the same as the z_b-axis. The angle, β_p is the sum of local twist, β_T and blade pitch angle, β, while φ is the inflow angle. Aerodynamic lift and drag forces, dF_L and dF_D are defined on the x_{ae} and y_{ae} axes, respectively (Ning 2013; Sahin 2018).

Using the above coordinate systems, the definition of turbine nacelle/rotor tilt angle, rotor pre-cone angle, etc. becomes possible in the HAWT model. Different from Ref (Manwell, McGowan, and Rogers 2009; Hansen 2008), here, BEM theory is implemented such that it calculates the aerodynamics of the turbine blades, separately. This allows the calculation of aerodynamic behavior for every blade individually due to each blade's varying attitude in time. To define aerodynamic calculations in the HAWT model, indices are used here to indicate which blade element section belongs to which blade of the turbine. The first lower case letter "i" is used to denote the blade such as 1st blade, 2nd blade, and 3rd blade, etc., while the second lower case letter, "j" represents the blade element section. In this chapter, only the required forces and moments are given in modeling. More information can be found in Ref(Sahin 2018). Therefore, considering BEM theory, employed aerodynamic corrections, particular coordinate systems, BEM-based HAWT model theory can be defined as follows.

Following Ref (Hansen 2008; Manwell, McGowan, and Rogers 2009), BEM theory calculates the angle of attack (AOA) of each blade element as

$$\alpha_{i,j} = \varphi_{i,j} - (\beta_{T_{i,j}} + \beta), \quad (1)$$

where φ represents the inflow angle, β_T is the twist angle, β is the blade pitch angle. The relationship between the inflow angle and the velocities at a blade element is as

$$\varphi_{i,j} = \tan^{-1}\frac{V_{bx_{i,j}}(1-a_{i,j})}{V_{by_{i,j}}(1+a'_{i,j})} \tag{2}$$

The above velocity components, V_{b_x} and V_{b_y} are calculated using equations (3) and (4). Here, they are obtained after a transformation of freestream wind speed, U from the wind-aligned to the blade-aligned coordinate system including the effect of blade rotation, $\Omega r cos\Phi$. Therefore, they include the effects of rotor precone angle, Φ, nacelle/rotor tilt angle θ, turbine yaw angle, Ψ, azimuth angle, Λ. The wind speed, U has u, v, and w velocity components at the wind-aligned coordinate system. Here, the wind is considered as blowing only in x_w-axis direction (Ning et al. 2015). Thus, not the velocity component, u, but other components, v, w are taken as zero. In addition, the x_i-axis is assumed to show the main wind direction and therefore, β_{wind} (Figure 3a) is neglected.

$$V_{bx_{i,j}} = u\big(cos\Phi_{i,j}cos\theta cos\Psi + sin\Phi_{i,j}cos\Lambda_{i,j}sin\theta cos\Psi + sin\Phi_{i,j}sin\Lambda_{i,j}sin\Psi\big) \tag{3}$$

$$V_{by_{i,j}} = u\big(sin\Lambda_{i,j}sin\theta cos\Psi - cos\Lambda_{i,j}sin\Psi\big) + \Omega r cos\Phi_{i,j} \tag{4}$$

The third velocity component, V_{b_z} is ignored since it is along the turbine blade and towards the blade tip, and therefore it does not contribute to the aerodynamic calculation. In equations (3) and (4), u is the freestream wind speed coming towards the turbine, Ω is the rotor speed, r is the local blade radius or the distance of each blade element to the rotor hub center. In equation (2), a and a' are the axial and tangential induction factors and are found by equations (5) and (6) through an iteration process (Hansen 2008).

$$a_{i,j} = \left(1 + \frac{4F_{i,j}\sin^2\varphi_{i,j}}{\sigma_{i,j}C_{n_{i,j}}}\right)^{-1} \tag{5}$$

$$a'_{i,j} = \left(-1 + \frac{4F_{i,j}\sin\varphi_{i,j}\cos\varphi_{i,j}}{\sigma_{i,j}C_{t_{i,j}}}\right)^{-1} \tag{6}$$

where C_n and C_t are normal and tangential force coefficients (i.e., when no precone angle is defined) and are calculated by equations (7) and (8), respectively (Hansen 2008).

$$C_{n_{i,j}} = (C_{l_{i,j}} \cos \varphi_{i,j} + C_{d_{i,j}} \sin \varphi_{i,j}) \tag{7}$$

$$C_{t_{i,j}} = (C_{l_{i,j}} \sin \varphi_{i,j} - C_{d_{i,j}} \cos \varphi_{i,j}) \tag{8}$$

The solidity, σ of each blade element is determined as follows:

$$\sigma_{i,j} = \frac{B c_{i,j}}{2\pi r_{i,j}} \tag{9}$$

The aerodynamic loss due to the vortices shed from the blade tips is modeled using the Prandtl tip loss factor formula (10). For the hub vortices, a similar formulation (11) can be employed. Therefore, the total loss effect is calculated through equation (12)(Moriarty and Hansen 2005).

$$F_{tip_{i,j}} = \frac{2}{\pi} cos^{-1}\left(e^{-\frac{B}{2}\frac{R-r_{i,j}}{r\, sin\, \varphi}}\right) \tag{10}$$

$$F_{hub_{i,j}} = \frac{2}{\pi} cos^{-1}(e^{-\frac{B}{2}\frac{r_{i,j}-R_{hub}}{r_{i,j} sin\, \varphi_{i,j}}}) \tag{11}$$

$$F_{i,j} = F_{hub_{i,j}} F_{tip_{i,j}} \tag{12}$$

where $B, c, R,$ and R_{hub} are rotor blade number, elemental chord length, rotor, and rotor hub radius, respectively.

There are various models to represent the turbine behavior in the turbulent wake state operation(Glauert 1935; Buhl 2005; Hansen 2008). Here, Spera's correction formula (13) is used to correct the performance prediction of BEM theory for the turbine operation in the turbulent wake state(Hansen 2008; Branlard 2011).

$$a_{i,j} = \frac{1}{2}\left[2 + H(1 - 2a_c) - \sqrt{(H(1 - 2a_c) + 2)^2 + 4(Ha_c^2 - 1)}\right] \tag{13}$$

where a_c is the critical axial induction factor. H is calculated by equation (14).

$$H = \frac{4F_{i,j} \sin^2 \varphi_{i,j}}{\sigma_{i,j} C_{n_{i,j}}} \tag{14}$$

In addition, when the turbine operates with the skewed wake flow. The predictions of the BEM theory is corrected by Pitt and Peters formula (15) (Pitt and Peters 1981; Moriarty and Hansen 2005), which corrects the axial induction factor distribution throughout each blade span, and Burton formula (16) (Burton et al. 2001; Moriarty and Hansen 2005), which calculates the skewed wake angle.

$$a_{skew_{i,j}} = a_{i,j}(1 + \frac{15\pi}{32}\frac{r}{R}\tan\frac{X}{2}\sin\Lambda_{i,j}) \tag{15}$$

$$X = (0.6a + 1)\Psi \tag{16}$$

where X, Λ, and Ψ (Figure 4) are respectively the skewed wake, azimuth, and turbine yaw angles.

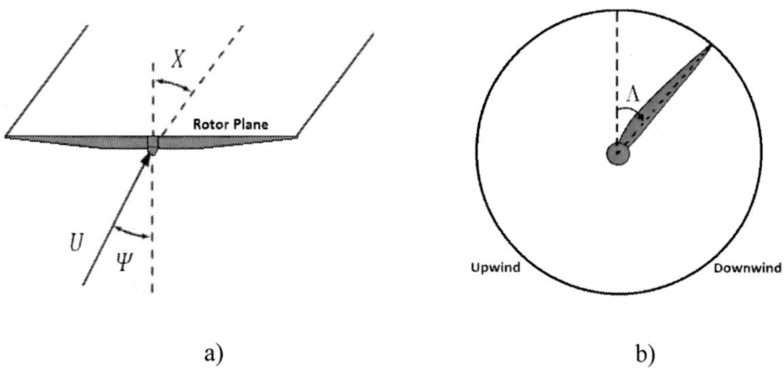

Figure 4. Coordinates a) Top view, b) Front view (Sahin 2018).

The effective air velocity, W, which each blade element exposes, can be calculated by equation 17 using V_{b_x} ve V_{b_y} velocity components in equation (3) and (4), respectively(Ning et al. 2015):

$$W_{i,j} = \sqrt{(V_{b_{x_{i,j}}}(1 - a_{i,j}))^2 + (V_{b_{y_{i,j}}}(1 + a'_{i,j}))^2} \tag{17}$$

Inside the iteration loop for calculating the induction factors, the following are carried out: axial and tangential induction factors, a and a' are first initialized with zeros, inflow angle, φ, AOA, or α, tip loss, hub loss, F_{tip} and F_{hub} , total loss, F are calculated through the equations (2), (1), (10), (11), and (12), respectively. Aerodynamic lift and drag coefficients, C_l and C_d are selected according to the estimated AOA of each blade element through an interpolation. Later, normal and tangential force coefficients, C_n and C_t, are obtained by equations (7) and (8). Solidity, σ of each blade element is found by equation (9). Axial and tangential induction factors, a and a' are calculated by equations (5) and (6). If the turbine operates in the turbulent wake state, i.e., an estimation of axial induction factor larger than the critical value, a_c, the axial induction factor is corrected through the Spera's correction formula (13). After satisfying the convergence criteria(s) for the iteration loop, both induction factors are obtained. Skewed wake correction is applied to the axial induction factor through equations (15) and (16) if the turbine operates with a yaw or tilt angle (or both). Here, the skewed wake phenomenon (Figure 4) is illustrated considering the yawed turbine operation. Thus, yaw angle, Ψ is used in equation (16), and $\sin\Lambda$ is used in equation (15). If the skewed wake occurs due to tilted nacelle or rotor, nacelle tilt angle, θ is to be in equation (16), and $\cos\Lambda$ is to be used in equation (15). To account for the skewed wake flow due to both nacelle tilt and yaw angles, more modification is required in modeling. This application is carried out outside the iteration loop. All of the above processes are employed to calculate induced velocities, aV_{b_x} and $a'V_{b_y}$ first and then the velocity components, $V_{b_x}(1-a)$ and $V_{b_y}(1+a')$ at the blade-aligned coordinate system. Later, using these velocity components, the effective wind speed, W (17) seen by each blade element is obtained to calculate the elemental aerodynamic lift force, dF_L (18), and drag force, dF_D (19) respectively in the x_{ae}-axis and y_{ae}-axis of aerodynamic coordinate system.

$$dF_{L_{i,j}} = \frac{1}{2}\rho W_{i,j}^2 C_{l_{i,j}} c_{i,j} dr_{i,j} \tag{18}$$

$$dF_{D_{i,j}} = \frac{1}{2}\rho W_{i,j}^2 C_{d_{i,j}} c_{i,j} dr_{i,j} \tag{19}$$

There is no aerodynamic force along the z_{ae}-axis. Here, the aerodynamic torsion moments are also neglected. Through the transformation of the above elemental lift and drag forces from the aerodynamic coordinate system to the blade-aligned coordinate system, the normal and tangential forces (i.e., no precone angle) are obtained in the blade-aligned coordinate system as

$$dF_{N_{i,j}}=dF_{L_{i,j}}\cos\varphi_{i,j}+\ dF_{D_{i,j}}\sin\varphi_{i,j} \tag{20}$$

$$dF_{T_{i,j}}=dF_{L_{i,j}}\sin\varphi_{i,j}-\ \ dF_{D_{i,j}}\cos\varphi_{i,j} \tag{21}$$

Afterwards, considering the position or local radius of each blade element with respect to the hub center, the moments can be calculated. All these above forces and moments are later transformed to the hub-aligned coordinate systems. Summation of all the elemental forces and moments are throughout every turbine blade span, the total turbine rotor forces and moments in x_h, y_h, z_h axes are obtained. Here, only the required elemental force (22) and moment (23) in the x_h-axis, which respectively create turbine aerodynamic thrust and rotor torque, are presented.

$$F_{h_{x_{i,j}}}=dF_{N_{i,j}}\cos\Phi_{i,j} \tag{22}$$

$$M_{h_{x_{i,j}}}=dF_{T_{i,j}}r_{i,j} \tag{23}$$

When the above elemental forces and moments are summed throughout all the turbine blades, the total aerodynamic thrust, F_t (24) and rotor torque, τ_{aero} (25) are respectively calculated as

$$F_t=\textstyle\sum_i^B\sum_j^S F_{hx_{i,j}} \tag{24}$$

$$\tau_{aero}=\textstyle\sum_i^B\sum_j^S M_{hx_{i,j}} \tag{25}$$

where B is the number of turbine blades, S is the number of blade elements. Using the equation (26), turbine power, P, is also obtained taking into account the rotor speed, Ω.

$$P=\tau_{aero}\Omega \tag{26}$$

Eventually, the rotor precone angle, Φ reduces the rotor disk area, while the tilt, θ and yaw, Ψ angles of the turbine affect the amount of wind flow passing through the turbine rotor disk. Thus, a turbine with a tilt or a yaw angle towards a freestream wind flow is subjected to a wind speed component normal to the rotor disk rather than the actual wind speed. This situation affects the power of wind flow passing through the rotor and the dynamic pressure force applied to the turbine rotor. Therefore, these structural angles must be included in the calculation of turbine TSR, λ, power coefficient, C_p and thrust coefficient C_T. Then, the TSR of the turbine with precone, tilt, and yaw angles can be obtained as follows (Sahin 2018).

$$\lambda = \frac{\Omega R \cos \Phi}{u \cos \theta \cos \Psi} \tag{27}$$

The power available in the wind can be obtained by equation (28),

$$P_{wind} = \frac{1}{2}\rho(u \cos \theta \cos \Psi)^3 \pi (R \cos \Phi)^2 \tag{28}$$

The dynamic force due to the wind passing through the rotor is calculated by equation (29),

$$F_{dynamics} = \frac{1}{2}\rho(u \cos \theta \cos \Psi)^2 (R \cos \Phi)^2 \tag{29}$$

Hence, power and thrust coefficients (C_p and C_T) are then found respectively,

$$C_p = \frac{P}{\frac{1}{2}\rho(u\cos\theta\cos\Psi)^3 \pi (R\cos\Phi)^2} \tag{30}$$

$$C_T = \frac{T}{\frac{1}{2}\rho(u\cos\theta\cos\Psi)^2 \pi (R\cos\Phi)^2} \tag{31}$$

Here, the HAWT system is considered as a one-mass model(Figure 5), i.e., a turbine rotor (a hub with blades) connected to a perfectly stiff and frictionless shaft mechanism that includes a Low Speed Shaft (LSS) and a High Speed Shaft (HSS).

In the HAWT system, the rotor aerodynamic torque, τ_{aero} acts opposite to the electromagnetic torque, τ_{gen} of the turbine electrical generator. Therefore, using the Newton second law of motion, a HAWT system can be represented by the following first-order differential equation (32) as

$$J_t \dot{\Omega} = \tau_{aero} - \tau_{gen} \tag{32}$$

$$J_t = J_r + N_{gear}^2 J_{gen} \tag{33}$$

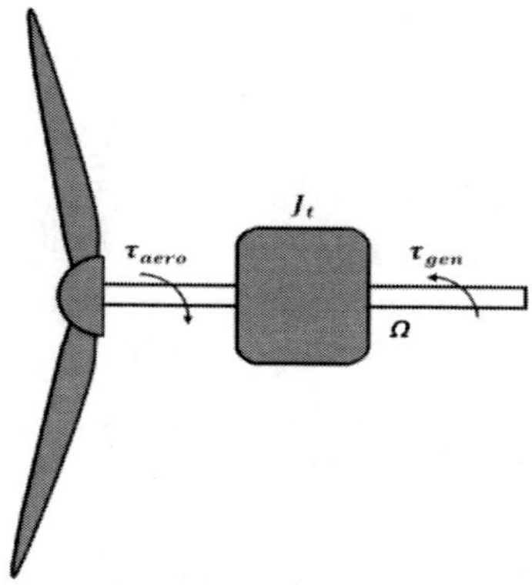

Figure 5. One-mass wind turbine model (Sahin 2018).

where J_t is the total inertia of the HAWT system obtained from the rotor inertia, J_r and the generator inertia, J_{gen} with the effect of gearbox ratio, N_{gear}, Ω is the rotor speed, τ_{aero} is the previously obtained turbine aerodynamic rotor torque, while τ_{gen} is the generator electromagnetic torque on the rotor shaft, i.e., LSS of the gearbox (Sahin 2018). Therefore, it includes the effect of the gearbox ratio. This is a nonlinear HAWT model due to both aerodynamic and generator torques. The generator electromagnetic torque, τ_{gen} is adjusted through a generator torque controller unit, i.e., power electronics on variable speed turbines. Since the theory defined above is the same as that of the MS Bladed Model. The rest of this chapter continues with the MS Bladed Model implementation, operation principle, and its validation.

MS Bladed Model and Its Validation

The MS Bladed Model has similarities to PROP code(Wilson and Lissaman 1974), PROPID (M. S. Selig 2012), Wt_Perf (Platt and Buhl 2012), and AeroDyn (Moriarty and Hansen 2005) software in the sense of aerodynamic calculations. Figure 6 shows the flowchart of the MS Bladed Model (Sahin 2018), which is developed by the corresponding author of this chapter using the above theory in MATLAB and Simulink programs. The model consists of three parts: Part I, II, and III. Part I and II, which are used together, are implemented in MATLAB to calculate turbine aerodynamics. The model is validated with field data of NREL experimental turbines. Afterwards, the validated model was moved into a Simulink block. This block and Part III are combined in the Simulink program (Sahin 2018). In the flowchart, shown in blue color is Part I. It includes wind, wind turbine, and rotor properties such as wind speed, air density, blade number, nacelle tilt angle, rotor precone angle, spanwise blade twist, and cord distributions, etc. Part II is seen in black color. It deals with the aerodynamic calculations based on BEM theory along with aerodynamic corrections such as turbulent wake state, skewed wake, etc. Part III, depicted in orange color, focuses on the model dynamisation, using both aerodynamic rotor and electrical generator torques, including the effect of gearbox ratio, and their respective inertias along with the rotor speed (Sahin 2018).

The MS Bladed Model works as follows: Introducing the wind (level, density, etc.), and turbine properties (rotor pre-cone and nacelle tilt angles, blade chord and twist distributions, etc.), an iteration process – along with the skewed wake correction - calculates the elemental forces and moments at the rotor hub. It then sums those elemental loads through each turbine blade span to calculate total loads, i.e., forces and moments. The force and moment in the direction of the x-axis of the hub-aligned coordinate system are the turbine total thrust and torque, respectively. The torque is later used as the counteracting generator torque considering the gearbox ratio to construct the MS Bladed Model (Sahin 2018).

From the MS Bladed Model, the users can obtain individual blade information such as torque and thrust. It is also possible to obtain spanwise distributions of axially and tangentially induced velocities, inflow angles, angles of attacks, hub and tip loss factors, total loss factor, axial and tangential induction factors, lift and drag forces and their respective coefficients, tangential and normal forces to the $y_b - z_b$ plane, etc. (Sahin 2018).

Validation of the BEM-based turbine model has to be realized first to use it for turbine performance predictions, turbine simulations, controller designs, new algorithm implementations, and so on. Here, as an example, validation of the MS Bladed Model, whose calculations are based on the theory defined above, are presented. Before elaborating on the details about the model validation process, a summary of the United States National Renewable Energy Laboratory (or NREL) is particularly felt useful since test data of the experimental turbines used for model validation in this subchapter are generated by this laboratory.

NREL, located in Golden/Colorado, is a part of the US Department of Energy Office (DOE) for renewable energy and energy efficiency. The laboratory started working in 1977 as a Solar Energy Research Institute (SERI) and was renamed, later in 1991 as NREL. A part of NREL is the National Wind Technology Center (NWTC), which realizes applied researches and various tests with industrial partners, ranging from small to large turbine manufacturers (Johnson et al. 2006). NREL has been utilizing the NWTC to carry out different turbine tests. NREL Phase II and Phase III experimental turbine tests are few examples of these tests (Schepers et al. 2002).

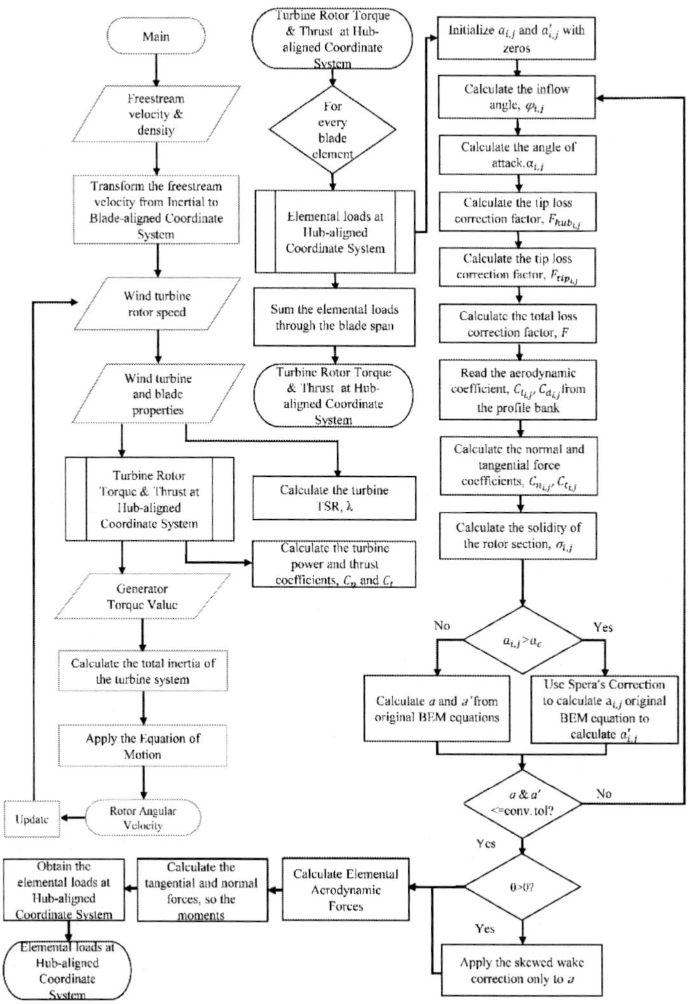

Figure 6. MS Bladed Model Flowchart (Sahin 2018).

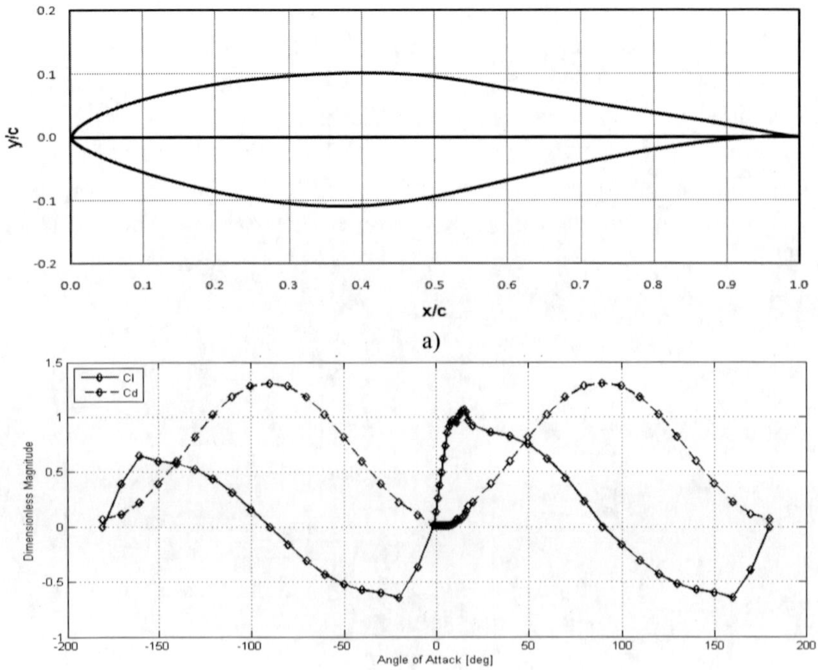

a)

b)

Figure 7. NREL S809 airfoil, a) Airfoil shape and coordinates (NREL 2017), b) Extrapolated aerodynamic data of S809 at Re number of $1{,}0.10^6$, $C_{d,90}$ of 1.308 (Sahin 2018).

A decade after its operational start, NREL began an extensive test program referred to as Combined Experiment in 1987 to investigate the complex 3D aerodynamics of wind turbines. The tests with different turbine configurations were initiated as Phase I and Phase II. Afterwards, these tests were extended to include Phase III, IV, V, and VI, under a new name referred to as Unsteady Aerodynamic Experiments (UAE). In all these tests, many different modifications to the test setup were conducted according to the needs. Different blade sets consisting of NREL S809 airfoil (Figure 7a) were utilized through all the experiments. These are constant chord/untwisted blades in Phase I (Butterfield, Musial, and Simms 1992) and II, constant chord/optimally twisted blades in Phase III, IV (Simms et al. 1999), and V (Hand, Simms, Fingersh, Jager, and Cotrell 2001), lastly tapered and nonlinearly twisted blades in Phase VI (Hand, Simms, Fingersh, Jager, Cotrell, et al. 2001).

For the validation of the MS Bladed Model, the test results of Phase II and III experimental turbines are used in this chapter. More validation cases with other NREL turbines such as Phase VI experimental turbine and NREL 5 MW land-based turbine are available in Ref (Sahin 2018). There are common properties of NREL Phase I to Phase VI turbines such as the usage of same NREL S809 airfoil for all the blade sets.

For validation purposes, the blades of all the turbines are divided into 20 elements. In the MS Bladed Model, axial and tangential induction limitations, error tolerance, or convergence criterion are kept the same for the power predictions of both NREL Phase II and III experimental turbines. The critical axial induction factor,a_c is taken as 0.37.

For the NREL Phase II and III turbines, the experimental aerodynamic data between the AOAs of -1 and 18 degrees measured at Reynolds number of $1{,}0.10^6$ are utilized for the predictions of power output (Somers 1997; M. Selig 2017). Outside this AOA range, the aerodynamic data are obtained here using the Airfoilprep program (NREL 2014). This program extrapolates this limited amount of experimental aerodynamic data from -180 to 180 degrees (Figure 7b). During the extrapolation, the aspect ratio, or AR, the ratio of blade length over chord length at 80% radius, is chosen as 11 (Tangler and Kocurek 2005). Afterwards, the extrapolated aerodynamic data are used within the MS Bladed Model.

For the Phase II and III experiments, NREL utilized the Grumman Wind Stream 33 turbine. It is a three-bladed downwind, stall-regulated HAWT with a rotor diameter of 10 m. The rated power of the turbine is about 20 kW and has a fixed speed of 72 rpm. Table 1 shows the main specifications of this test turbine. The Grumman turbine has the properties of free-yaw and manual full-span pitch control. To employ the turbine for the above test phases, vital modifications to the turbine were applied according to the needs. For instance, it is instrumented to characterize rotating blade aerodynamics, structural response as well as atmospheric inflow conditions. However, the most dominant modification on the turbine was the change of Grumman blades with those of NREL blades made of the S809 profile (Simms et al. 1999). The S809 airfoil is depicted in Figure 7a. This airfoil has a low maximum lift, a minimal sensitivity of maximum lift to leading edge roughness as well as a low profile drag (Somers 1997) (Gerber et al. 2005). Additionally, this airfoil was utilized to design the NREL blades for all the test turbines, ranging from Phase I to Phase VI.

Table 1. Test turbine basic properties (Schepers et al. 2002)

Turbine Type	HAWT
Rotor Location	Downwind
Blade Number	3
Rotor Diameter	10.06 m
Root Extension	0.723 m
Rotor Type	Fixed Speed
Power Regulation	Stall-Regulated
Operational Rotor Speed	71.63 rpm
Tilt Angle	0 degrees
Precone Angle	3.25 degrees
Cut-in Wind Speed	6 m/s
Cut-out Wind Speed	NA (stall control)
Rated Power	19.8 kW

In the next pages, the MS Bladed Model is validated using the NREL Phase II, and III turbine experimental power data. Although the turbine blades are divided into 20 elements in the MS Bladed Model, the spanwise AOA distribution is given at the last 17 elements. This is because the effect of the first three elements is neglected as they correspond to the hub extension where there is no S809 airfoil.

NREL Phase II Wind Turbine

NREL Phase II is an experimental wind turbine with three blades. It has neither twisted nor tapered blades with a length of 5.03 m, and a constant chord length of 0.4572m throughout the span. The turbine does not have a nacelle tilt angle, but a precone angle of 3.25 degrees. Also, blades are fixed to the rotor hub, approximately 12 degrees. The operational rotor speed of the turbine is 72 rpm. The chord and twist distributions of the NREL Phase II turbine blades are given in Ref (Schepers et al. 2002).

Figure 8 shows the power prediction of the MS Bladed Model and NREL Phase II turbine test data as well as Ceyhan's numerical predictions at various freestream velocities (Ceyhan 2008). The solid black curve with square symbols is produced from the NREL test data (Schepers et al. 2002). The dashed and solid blue curves with diamond symbols are respectively the MS Bladed Model predictions for the Low Speed Shaft (LSS) mechanical rotor power and electrical generator power. The predictions are based on the extrapolated experimental aerodynamic data given in Figure 7b. The dashed red curve with hexagram symbols is Ceyhan's numerical prediction (Ceyhan 2008). As there is a slight difference between the experimental data used in this chapter and those of Ceyhan, there is a minor difference between the two predictions, particularly at high wind speeds (Sahin 2018).

Here, the performance prediction of the MS Bladed Model is given in terms of mechanical rotor power. However, NREL test data are presented in terms of electrical power. Due to friction in the gearbox, copper losses in the generator, and so on, there are some mechanical and electrical losses in the turbine system(Sahin 2018). In Ref(Schepers et al. 2002), it is stated that the NREL test turbine has an efficiency of fairly fixed and approximately 78%. However, the MS Bladed Model considers a 100% efficiency in the turbine system. Therefore, when the efficiency of 78% is taken into consideration, the solid blue curve in Figure 8 represents the predicted power output of the MS Bladed Model.

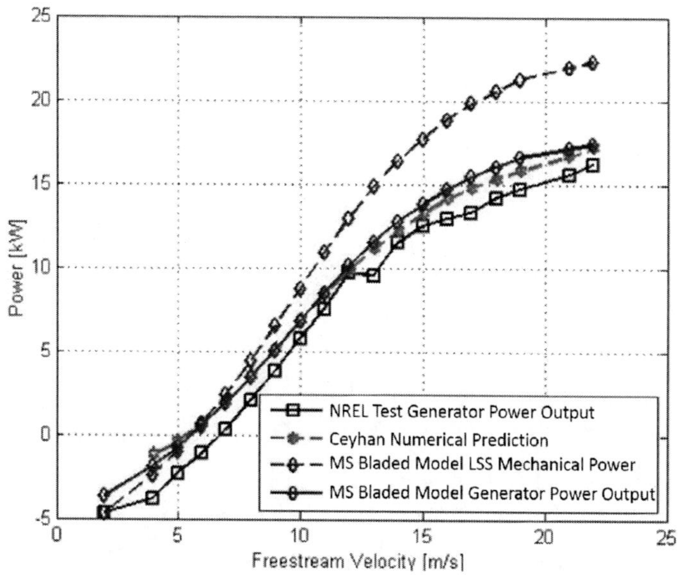

Figure 8. Power predictions for NREL Phase II turbine (Sahin 2018).

Starting at around 8 m/s to high wind speeds, the MS Bladed Model predicts the turbine performance using the extrapolated aerodynamic data. Figure 9 shows the spanwise AOA distributions at wind speeds of 8 and 21 m/s. At the wind speed of 8 m/s, only the first three blade elements operate with an AOA of above 18 degrees (Figure 9a), which corresponds to the extrapolated aerodynamic data. However, at the wind speed of 21 m/s, the first fourteen elements operate with AOAs of above 18 degrees (Figure 9b), meaning that few blade elements operate with the experimental data (Sahin 2018).

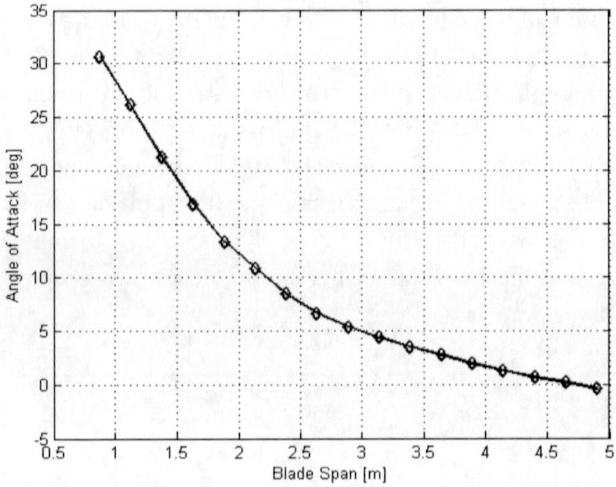

a)

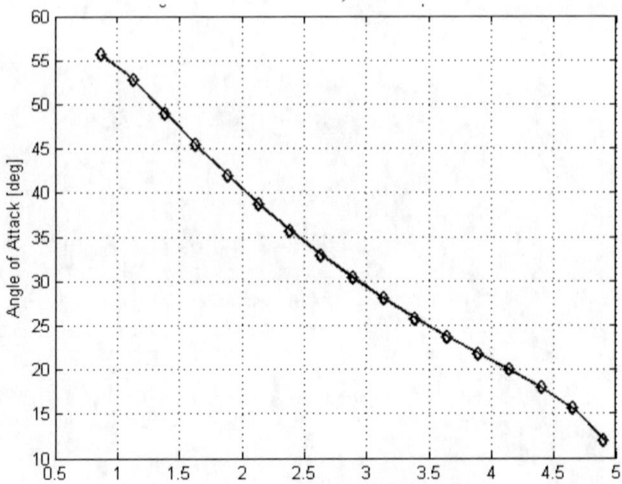

b)

Figure 9. Spanwise AOA distributions, a) 8 m/s wind speed, b) 21 m/s wind speed (Sahin 2018).

As the wind speed increases further, beginning from the blade root, most blade sections start operating with the extrapolated aerodynamic data. At high wind speeds, such as the case at the wind speed of 21 m/s (Figure 9), the estimated performance depends mostly on the extrapolated data. As seen in Figure 8 for the NREL Phase II configuration, which employs untwisted and untapered blades, the power predictions of the MS Bladed Model with the extrapolated data, are quite satisfactory against the NREL test data(Sahin 2018).

NREL Phase III Wind Turbine

NREL Phase III turbine is also an experimental wind turbine. It has twisted, but not tapered three blades with a length of 5.03 m and a constant chord of 0.4572 m through the span. The turbine does not have a tilt angle, but a precone angle of 3.25 degrees. Also, the blades are fixed to the rotor hub with a pitch angle of approximately 3 degrees. The twist and chord distributions of the NREL Phase III turbine blades are found in Ref (Schepers et al. 2002). The operational rotor speed of the turbine is the same as the Phase II test turbine, i.e., 72 rpm.

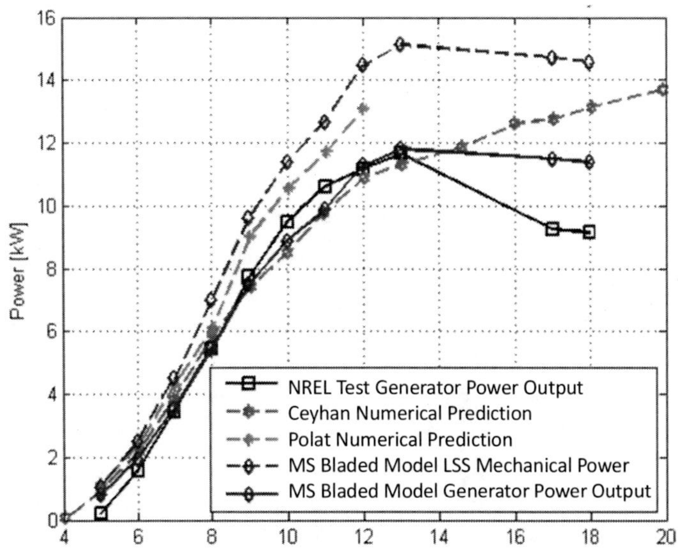

Figure 10. Power predictions for NREL Phase III turbine (Sahin 2018).

Figure 10 shows four different performance predictions and the NREL test data (Schepers et al. 2002). The NREL test data are given with the solid black curve with square symbols. Ceyhan's numerical prediction(Ceyhan 2008) is the red dashed curve with hexagram symbols, while Polat's (Polat 2011) numerical prediction is shown by the magenta-colored curve with stars. Both Ref(Ceyhan 2008) and Ref (Polat 2011) utilize the same experimental aerodynamic data for power predictions. As in the Phase II case, the dashed and solid blue curves with diamond symbols in Figure 10 are the MS Bladed Model predictions for the rotor mechanical power and generator power, respectively. The same efficiency, 78% is still valid for Phase III experimental turbine because the difference on the turbine is only the blade sets. Thus, the solid blue curve with diamond symbols represents the electrical power output estimated by the MS Bladed Model.

Although the same extrapolated aerodynamic data (Figure 7b) are utilized for the NREL Phase III turbine power predictions, there appear some differences between the currently predicted generator power of the MS Bladed Model and the NREL test data. This situation is particularly visible when the wind speed exceeds the wind speed of approximately 9 m/s. Nevertheless, most of the predicted powers of the MS Bladed Model are almost the same as the NREL test data, except at 5, 10, and 11 m/s wind speeds as well as the wind speeds beyond 13 m/s. Thus, the possible reasons for these differences in power outputs may be explained as follows (Sahin 2018).

Figure 11a shows the spanwise AOA distribution at the wind speed of 5 m/s. The first five inner blade elements operate with negative AOAs using the extrapolated aerodynamic data. Therefore, the MS Bladed Model predicts a different power output from the NREL power test data. However, at the wind speed of 6 m/s (Figure 11b), only the first blade element uses the extrapolated aerodynamic data. Therefore, the MS Bladed Model produces almost the same power output as the NREL test data (Sahin 2018).

The MS Bladed Model predictions and the NREL test data are almost the same between the wind speeds of 6 and 9 m/s (Figure 10) since all the operational AOAs of all blade sections fall below the critical AOA of 9.22 degrees (Sahin 2018). For instance, Figure 12 shows the spanwise AOA values at wind speeds of 8 and 9 m/s, which are all below critical value of 9.22 degrees.

However, between the wind speeds of 10 and 11 m/s, the predicted power values are lower than the test data. Figure 13 shows the spanwise AOA distributions at these wind speeds. As seen in both figures, most blade elements operate with AOAs higher than the critical value of 9.22 degrees.

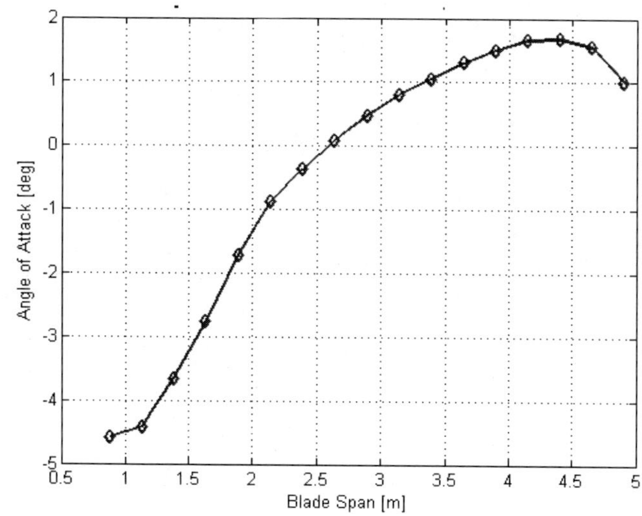

a)

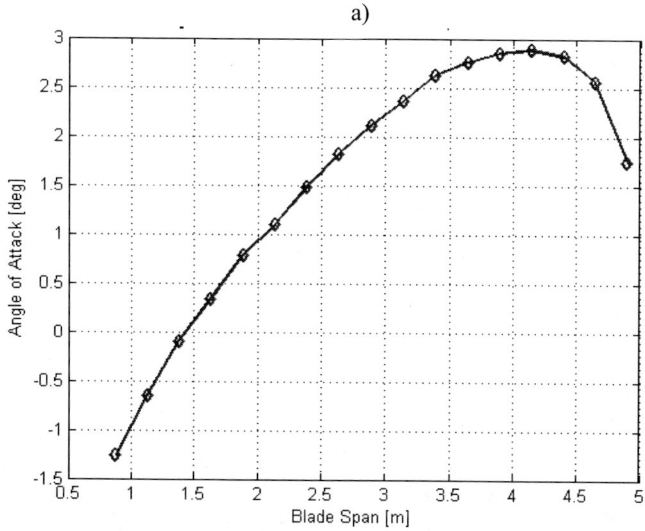

b)

Figure 11. Spanwise AOA distributions, a) 5 m/s, b) 6 m/s (Sahin 2018).

Thus, these blade elements work within the stall region. It is expected that the complex 3D stall delay effect, which occurs due to the blade rotation and changes the aerodynamic airfoil data through the blade span, is the reason for this difference. Therefore, the airflow over these blade elements is probably still attached to the blade surface. This causes the turbine blades to produce different lift and drag data than the measured 2D wind tunnel data, which increases turbine power output (Sahin 2018).

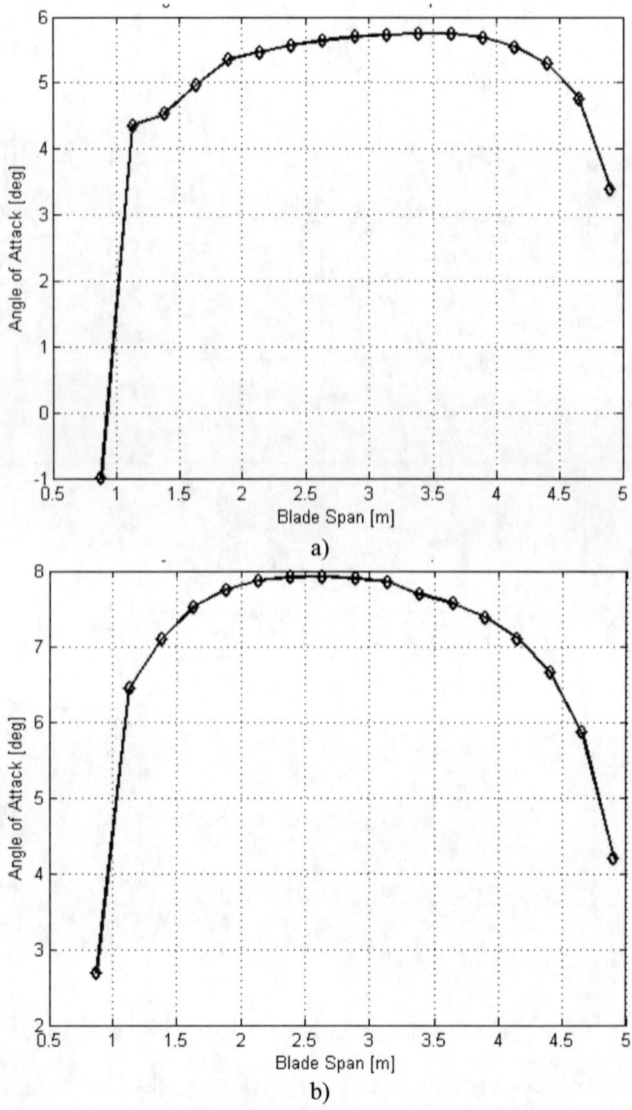

Figure 12. Spanwise AOA distributions, a) 8 m/s, b) 9 m/s (Sahin 2018).

Figure 14a and b show the spanwise AOA distributions at wind speeds of 12 and 13 m/s, respectively. From the figures, it is clear that the turbine operates mostly with the measured 2D wind tunnel data at these two wind speeds, namely within the stall region of the S809 airfoil. However, even in the stall conditions, the predicted power output is almost the same as the NREL test data. Thus, at these wind speeds, the total effect of 3D aerodynamics has the same total effect of 2D tunnel data along the blade span (Sahin 2018).

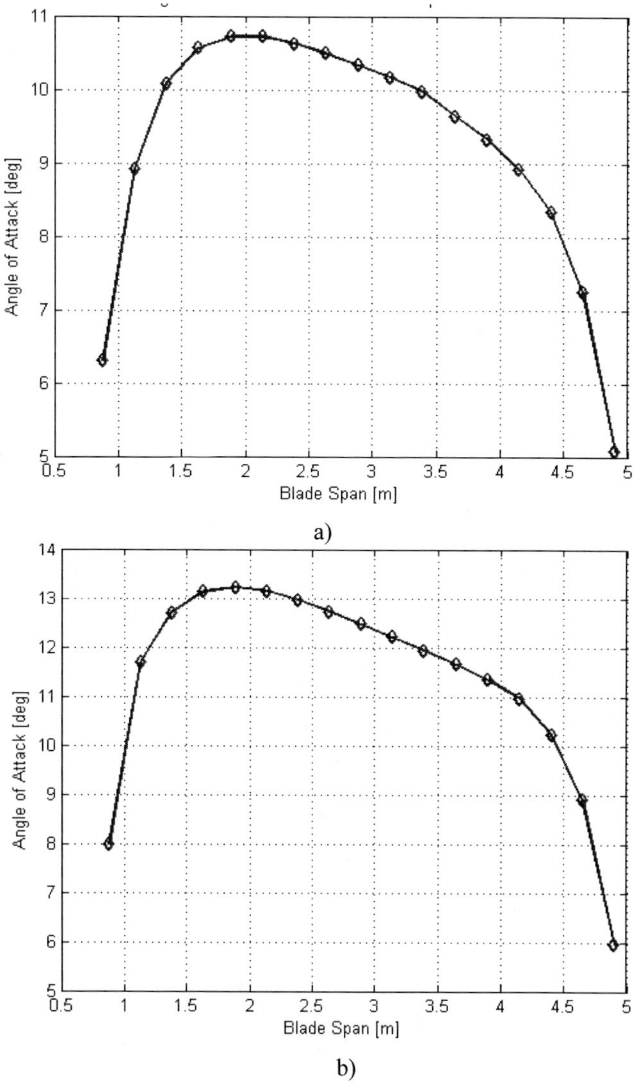

Figure 13. Spanwise AOA distributions, a) 10 m/s, b) 11 m/s (Sahin 2018).

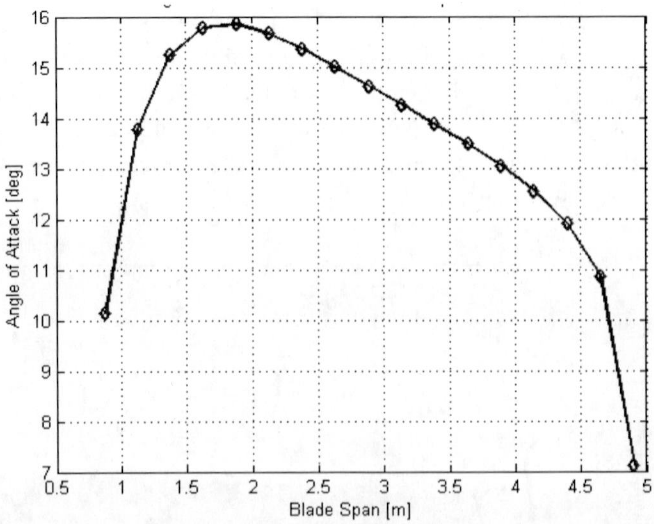

a)

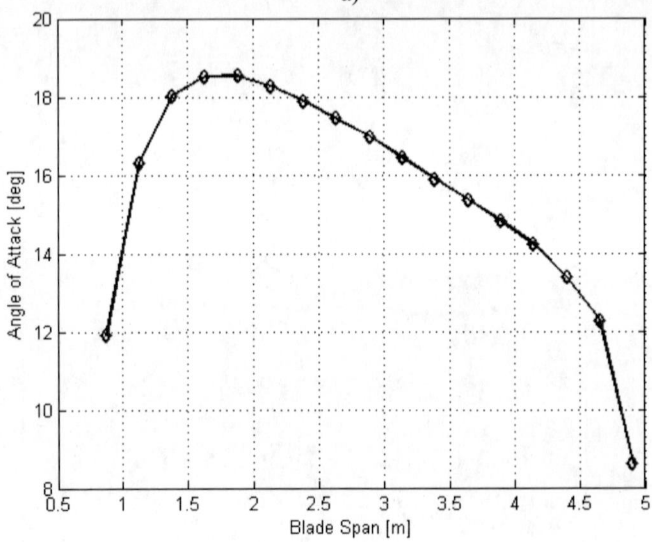

b)

Figure 14. Spanwise AOA distributions, a) 12 m/s, b) 13 m/s (Sahin 2018).

The spanwise AOA distributions at wind speeds of 17 and 18 m/s are shown in Figure 15a and b, respectively. The differences between the predicted powers and NREL test data become larger although the same extrapolated airfoil data (Figure 7b) are employed as in the case of the Phase II turbine.

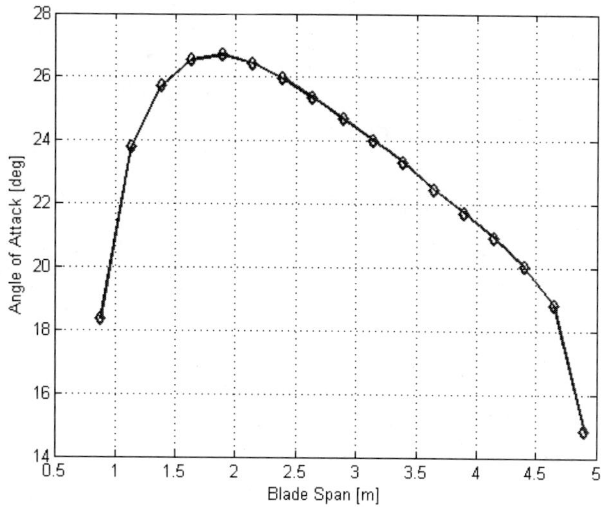

a)

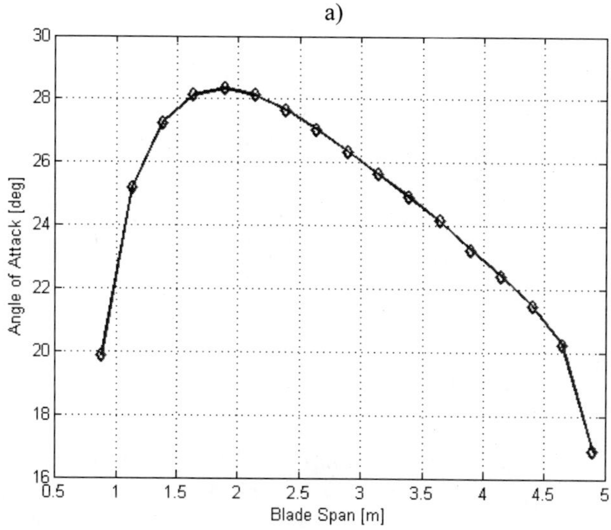

b)

Figure 15. Spanwise AOA distributions, a) 17 m/s, b) 18 m/s (Sahin 2018).

The possible reasons for this situation may be explained as follows: The extrapolated data at high wind speeds are not accurate enough to predict the turbine power output due to the highly twisted blades. At these wind speeds, all the blade elements, except the last one, operate under AOAs that utilize the extrapolated data. A similar reason is noted by Ref (Ceyhan 2008) as follows.

The local AOA values are larger than the stall angle. Thus, this decreases the reliability of the aerodynamic data at high wind speeds.

Adopted settings for the convergence tolerance and whether to include the drag coefficient into the calculation of the induction factors or not are expected as the other factors affecting the predicted performance of the MS Bladed Model. Nevertheless, the obtained results are eventually quite satisfactory for the validation of the MS Bladed Model.

Conclusion

In this chapter, modeling of rigidly structured Horizontal Axis Wind Turbines (or HAWTs) is covered. The HAWT modeling is defined based on the well-known Blade Element Momentum (BEM) theory. Also, various aerodynamic correction models for the turbulent wake state operation, skewed wake rotation, etc. are explained with the reasons of usage in modeling. Particular coordinate systems are defined and illustrated. These coordinate systems are used for fixed or time-varying structural angles such as nacelle tilt angle, precone angle, yaw angle, blades' azimuth angles, etc. The turbine model has the properties of individual and collective blade pitching, nacelle yawing, etc. The MS Bladed Model, which uses the same theoretical approach defined in this chapter, is given as an example turbine model. The implementation and operation principle of the model are defined. Validation of the MS Bladed Model is carried out in detail using published data of NREL experimental test turbines such as Phase II and III. Also, other numerical predictions for these turbines are presented. Lastly, it may be concluded that the MS Bladed Model is capable of predicting the performance of wind turbines as accurately as possible depending on the aerodynamic data utilized for turbine blade airfoils.

Acknowledgment

The corresponding author is very grateful to Dr. Jason Jonkman from the US National Renewable Energy Laboratory (NREL) for his valuable guidance, comments, sharing experiences during the development of the MS Bladed Simulation Model.

References

Branlard, Emmanuel. (2011). "Wind Turbine Tip-Loss Corrections." Master's Thesis, Technical University of Denmark, DTU.

Buhl, M. L. (2005). "A New Empirical Relationship between Thrust Coefficient and Induction Factor for the Turbulent Windmill State A New Empirical Relationship between Thrust Coefficient and Induction Factor for the Turbulent Windmill State." *Technical Report NREL/TP-500-36834*.

Burton, T., D. Sharpe, N. Jenkins, and E. Bossanyi. (2001). *Wind Energy Handbook*. West Sussex, England: John Wiley & Sons, Ltd.

Butterfield, C. P., W. P. Musial, and D. A. Simms. (1992). "Combined Experiment Final Report- Phase I, NREL/TP-257-4655, Golden, Colorado: National Renewable Energy Laboratory."

Ceyhan, Ö. (2008). "Aerodynamic Design and Optimization of Horizontal Axis Wind Turbines by Using BEM Theory and Genetic Algorithm." Middle East Technical University.

Erich, H. (2006). *Wind Turbines-Fundamentals, Tecnologies, Application, Economics. Spectrum*. 2nd Edi. Germany.

Gerber, B. S., J. L. Tangler, E. P. N. Duque, and J. D. Kocurek. (2005). "Peak and Post-Peak Power Aerodynamics from Phase VI NASA Ames Wind Turbine Data" 127 (May): 192–99. https://doi.org/10.1115/1.1862260.

Glauert, H. (1935). *"Airplane Propellers."* Aerodynamic Theory (W. F. Durand, ed.), Div. L, Chapter XI. Berlin:Springer Verlag.

GWEC. (2019). "GWEC | Global Wind Report 2019."

GWEC. (2021). "GWEC | Global Wind Report 2021."

Hand, M. M., D. A. Simms, L. J. Fingersh, D. W. Jager, and J. R. Cotrell. (2001). "Unsteady Aerodynamics Experiment Phase V: Test Configuration and Available Data Campaigns, NREL/TP-500-29491, Golden, Colorado: National Renewable Energy Laboratory."

Hand, M. M., D. A. Simms, L. J. Fingersh, D. W. Jager, J. R. Cotrell, S. Schreck, and S. M. Larwood. (2001). "Unsteady Aerodynamics Experiment Phase VI : Wind Tunnel Test Configurations and Available Data Campaigns, NREL/TP-500-29955, Golden, Colorado: National Renewable Energy Laboratory." http://www.osti.gov/energycitations/product.biblio.jsp?osti_id=15000240.

Hansen, M. O. L. (2008). *Aerodynamics of Wind Turbines*. 2th Edit. Earthscan.

Johnson, K. E., L. Y. Pao, M. J. Balas, and L. J . Fingersh. (2006). "Control of Variable Speed Wind Turbines-Standard and Adaptive Techniques for Maximizing Energy Capture." *IEEE Control Systems Magazine* 26 (3): 70–81.

Manwell, J. F., J. G. McGowan, and A. L. Rogers. (2009). *Wind Energy Explained, Theory, Design and Application*. Second Edi. John Wiley & Sons Ltd.

Moriarty, P J., and A C. Hansen. (2005). "AeroDyn Theory Manual, NREL/TP-500-36881, Golden, Colorado: National Renewable Energy Laboratory."

Ning, S. A. (2013). "CCBlade Documentation, Release 0.1.0."

Ning, S. A., G. Hayman, R. Damiani, and J. Jonkman. (2015). "Development and Validation of a New Blade Element Momentum Skewed-Wake Model within AeroDyn." In *AIAA Science and Technology Forum and Exposition 2015*. http://www.nrel.gov/docs/fy15osti/63217.pdf.

NREL. (2014). "Airfoilprep." (2014). https://nwtc.nrel.gov/AirFoilPrep.

NREL. (2017). "Nrel's S809 Airfoil Graphic and Coordinates." 2017. https://wind.nrel.gov/airfoils/shapes/S809_Shape.html.

Pao, L. Y., and K. E. Johnson. (2009). "A Tutorial on the Dynamics and Control of Wind Turbines and Wind Farms." In *2009 American Control Conference*, 2076–89. Hyatt Regency Riverfront, St. Louis, MO, USA. https://doi.org/10.1109/ACC.2009.5160195.

Pitt, D. M, and D. A Peters. (1981). "Theoretical Prediction of Dynamic-Inflow Derivatives." *Vertica, 5(1) March.*

Platt, A. D., and M. L. Buhl. (2012). "WT_Perf User Guide for Version 3.05.00."

Polat, O. (2011). "Genetic Algorithm Based Aerodynamic Shape Optimization of Wind Turbine Rotor Blades Using a 2-D Panel Method With a Boundary Layer Solver." Middle East Technical University.

Sahin, M. (2018). "Dynamic Modeling, Control and Adaptive Envelope Protection System for Horizontal Axis Wind Turbines, PhD Thesis, Department of Aerospace Engineering, METU, Ankara." Middle East Technical University.

Sahin, M., and I. Yavrucuk. (2017). "Dynamical Modelling of a Wind Turbine System with Precone and Tilt Angles." In *9th Ankara International Aerospace Conference*, 1–11. Ankara, Turkey.

Sahin, M., and I. Yavrucuk. (2019). "Performance Comparison of Two Turbine Blade Pitch Controller Design Methods Based on Equilibrium and Frozen Wake Assumptions." In *Ankara International Aerospace Conference, AIAC 2019*, 1–16. Ankara, Turkey.

Sahin, M., and İ. Yavrucuk. (2020). "Adaptive Envelope Protection Control for the below and above Rated Regions of Wind Turbines." *International Journal of Energy and Power Engineering* 14 (10): 275–83.

Schepers, J. G, A. J. Brand, A. Bruining, and et al. (2002). "Final Report of IEA Annex XVIII : 'Enhanced Field Rotor Aerodynamics Database', ECN-C-02-016."

Selig, M. (2017). "S809 Experimental Data from Delft Wind Tunnel." 2017. http://m-selig.ae.illinois.edu/propid/raj/wtprep/polardata-html/airfoil1.html.

Selig, M. S. (2012). "PROPID User Manual-Version 5.3.1."

Simms, D A, M M Hand, L J Fingersh, and D W Jager. (1999). "Unsteady Aerodynamics Experiment Phases II – IV Test Configurations and Available Data Campaigns, NREL/TP-500-25950." In . National Renewable Energy Laboratory (NREL), Golden, Colorado.

Singh, M, and S Santoso. (2011). "Dynamic Models for Wind Turbines and Wind Power Plants Dynam,NREL/SR-5500-52780, Golden, Colorado: National Renewable Energy Laboratory."

Somers, D. M. (1997). "Design and Experimental Results for the S809 Airfoil, NREL/SR-440-6918, Golden, Colorado: National Renewable Energy Laboratory," no. January 1997: 103.

Tangler, J., and J. D. Kocurek. (2005). "Wind Turbine Post-Stall Airfoil Performance Characteristics Guidelines for Blade-Element Momentum Methods." In *43rd AIAA Aerospace Sciences Meeting Meeting and Exhibit*. Reno, Nevada. https://doi.org/10.2514/6.2005-591.

Wilson, R. E., and P. B. S. Lissaman. (1974). *"Applied Aerodynamics of Wind Power Machines."*

Chapter 2

Wind Turbine Controller Design

Mustafa Şahin[1,*], PhD and İlkay Yavrucuk[2], PhD
[1]METU Center for Wind Energy Research, Middle East Technical University, Ankara, Turkey
[2]Department of Aerospace Engineering, Middle East Technical University, Ankara, Turkey

Abstract

This chapter focuses on the design of baseline turbine controllers. The design is carried out for the NREL 5 MW turbine using the MS (Mustafa Şahin) Bladed Wind Turbine Simulation Model. The controller design includes the generator torque and collective blade pitch controllers. It also presents the dynamic simulations of the controlled NREL 5 MW turbine system and investigates the steady-state turbine performance in terms of various turbine parameters such as power, thrust, blade pitch angle, rotor speed, and so on. Simulation results are used to show the performance of the MS Bladed Wind Turbine Model and the baseline controllers.

Keywords: wind turbine baseline controller designs, generator torque controller, blade pitch controller, The MS Bladed Wind Turbine Simulation Model, dynamic and steady-state turbine performance

---

*Corresponding Author's Email: musahin33@gmail.com; musahin@metu.edu.tr.

In: Wind Turbines
Editors: David Borge-Diez and Daniel Icaza
ISBN: 978-1-68507-974-1
© 2022 Nova Science Publishers, Inc.

Introduction

Designing controllers for wind turbines is an area for wind turbine research and development. Without controllers, wind turbines cannot produce the desired electrical output that is going to be connected to the electrical network. For instance, using advanced control strategies, the power efficiency of wind turbines can be increased, and the structural loadings can be decreased. Also, the turbines with advanced controllers can capture more energy from wind and possesses a longer life span (Pao and Johnson 2009).

In this chapter, controller design aspects for Horizontal Axis Wind Turbine (HAWT) systems are investigated using the MS (Mustafa Şahin) Bladed Wind Turbine Simulation Model or shortly the MS Bladed Model in the literature (Sahin 2018). The chapter is organized as follows. Firstly, brief information about HAWT technologies, control, and operating regions of Variable Speed Variable Pitch (VSVP) wind turbines are presented. Secondly, the MS Bladed Model is shortly explained. Afterward, the NREL 5 MW turbine is defined, briefly. Cp and Ct surfaces of the NREL 5 MW turbine are extracted using the MS Bladed Model. Next, using the properties of the NREL 5 MW turbine (Jonkman et al. 2009) within the MS Bladed Model, turbine baseline controllers such as generator torque controller and collective blade pitch controller are designed, respectively. Later, their dynamic simulations for the NREL 5 MW turbine are carried out through the use of the MS Bladed Model under step increasing wind speeds, ranging from Region II to Region III. Also, the steady-state turbine performance of the controlled 5 MW turbine is investigated in terms of different turbine parameters such as power, thrust, rotor speed, etc. Lastly, the conclusion is presented at the end of the chapter.

Horizontal Axis Wind Turbines

Horizontal Axis Wind Turbines or HAWTs are the widely used turbines in wind farms. They are also referred to as propeller turbines and are mostly seen with a three-bladed rotor placed on top of a tower to capture more energy from wind. Three blades are the best combination due to their balance, efficiency, and practicality, etc. Also, they may come up with upwind or downwind rotor configurations (Pao and Johnson 2009; Hemami 2012). For large-scale turbines, the rotor has a fixed precone angle and a nacelle has a tilt angle to put a clearance between the blade tips and the tower to prevent blade tip strikes.

Basically, in wind turbines, the turbine rotor captures the kinetic energy in the wind and turns that energy into rotational mechanic energy. This energy is then applied to an electrical generator through a shaft. When there is only one shaft for this purpose, this turbine is referred to as a direct-drive turbine. However, most HAWTs include a gearbox between the turbine rotor and the generator. Gearboxes are used to increase the rotational speed of the rotor shaft or low speed shaft (LSS) to drive the generator, which is connected to the high speed shaft (HSS) (Pao and Johnson 2009; Hemami 2012).

In wind turbine systems, passive and active control methods are used for power regulation. The passive method does not require an external power and occurs naturally, while the active method needs an external energy. For instance, stall regulation is a passive method and is utilized for stall regulated turbines whose blades are firmly attached to the rotor hub. This type of turbine does not require a sophisticated control system. In active control, however, turbine blades are actively pitched to stall or feather to regulate turbine power output. Also, turbines can be designed to operate at fixed or variable rotor speeds. Fixed speed turbines operate at almost constant rotor speed regardless of wind speed, whereas variable turbines operate at various rotor speeds. Variable speed operation allows turbines to have less structural loads, and more efficiency in generated power. Therefore, variable speed turbines are more efficient than fixed speed turbines. This may result in a 5% more annual energy captured by the variable speed turbines. Also, changing the blade pitch angles actively is another advantage for efficient power regulation. Therefore, within the last four decades, wind turbine design has reached technologies with or without a gearbox, with advanced power electronics, i.e, variable speed operation, enhanced aerodynamics, active control, and new drive train design from fixed speed operation, passive control with drive train with gearboxes (Letcher 2017).

Most modern turbines are variable speed and variable pitch turbines. In this chapter, the NREL 5 MW turbine (Jonkman et al. 2009), which falls into this category, is utilized. These types of turbines have control systems with several layers. These are referred to as the highest, middle, and lowest level controls. The highest level control, or supervisory control, decides the time to start or stop turbines depending on the level of wind speed. The wind speed that which turbines begin producing electricity is the cut-in wind speed. The wind speed, which is mostly 25 m/s, is the cut-out wind speed for most wind turbines. Just beyond this wind speed, turbines are stopped in order not to damage the turbine parts due to strong winds. However, when desired, turbines can be operated with extra control strategies, even beyond 25 m/s wind speed. Middle-level control is about the turbine's own control. It is known as the turbine operational control. The turbine generator torque, blade pitch, and yaw controllers fall into this control level. The generator torque control is carried out through a power electronics unit on variable speed wind turbines. This component adjusts the amount of electromagnetic torque to be supplied from the generator to the turbine rotor for turbine operation with the optimum turbine efficiency. The blade pitch control regulates the rotor speed at the rated value. Then, the rated power is achieved through the pitch controller. Yaw control, on the other hand, is utilized for directing the turbine nacelle and thereby the turbine rotor into the wind blowing direction. However, when compared to the generator torque and the blade pitch controls, yaw control is not such an interesting duty for control engineers. Therefore, this chapter focuses on the design and simulation of the generator torque and collective blade pitch controllers, only. The last one is the lowest control level. It is about internal generator control, actuator control, etc. This turbine control level must operate faster than the other turbine level controls (Johnson et al. 2006) for appropriate turbine performance.

In this chapter, the design of baseline controllers is carried out for the NREL 5 MW turbine (Jonkman et al. 2009). For this type of turbine, there are three different operating regions as Region I, II, and III. Below the cut-in wind speed is Region I at which wind turbines do not produce any electrical output owing to very low wind speeds. Within this region, wind speeds are not sufficient to generate electrical power even for turbines' own systems. Between the cut-in and cut-out wind speeds, there are two operating regions such as Region II and III. Region II is the region between the cut-in and rated wind speeds, while Region III is the region between the rated and cut-out wind speeds. The main goal in Region II is to maximize the energy capture of turbines as wind speeds in this region are low to produce the rated power. However, in Region III, turbines limit their powers to their rated value due to high wind speeds in order not to exceed certain mechanical and electrical loads, and thereby saving the turbine and its components. Also, Regions II and III are referred to as the turbine partial or below-rated and the turbine full load or above-rated regions, respectively. Some articles, such as Ref. (Oudah, Mohd, and Hameed 2014), take the cut–out wind speed into account as an extra turbine region, Region IV. Usually, above the cut-out wind speed, turbine operation is not allowed due to the extreme turbine loadings. Nevertheless, most modern utility-scale turbines operate even beyond the cut-out wind speed using one of the offline-shaped strategies such as ramp shaping, stepwise shaping, etc. (Stol and Fingersh 2004) (Johnson et al. 2006) (Merabet, Thongam, and Gu 2011) (Boukhezzar et al. 2007) (Hai et al. 2012) (Fischer and Shan 2013).

For example, Figure 1 illustrates turbine operating regions and their limits for a 5 MW wind turbine. The dashed red line is the rated power for the turbine. The blue curve shows the power available in wind. The green, on the other hand, is the controlled power curve of the turbine. As seen in Figure 1, In Region II, all the power available in wind cannot be turned into electrical energy due to the Betz limit and losses in turbine mechanical and electrical parts. These losses can be friction loss in the gearbox, copper loss in the generator, etc.

Between Region I and II, as well as Region II and III, there are also transitional regions referred to as Region $I^{1/2}$ and $II^{1/2}$, respectively. During the generator torque controller design, the transition from Region I to II or Region II to III is not carried out by a straightforward switching, but rather a dynamic way based on the generator (Petrović and Bottasso 2017) or rotor speed. The controllers in these transition regions are referred to as Region $I^{1/2}$ and $II^{1/2}$ transition controllers. Region I covers an operation from start-up to the cut-in wind speed. Just above the cut-in wind speed, turbines may not produce the maximum power possible (Region $II^{1/2}$), yet it is still considered to be operating in Region II (Oudah, Mohd, and Hameed 2014).

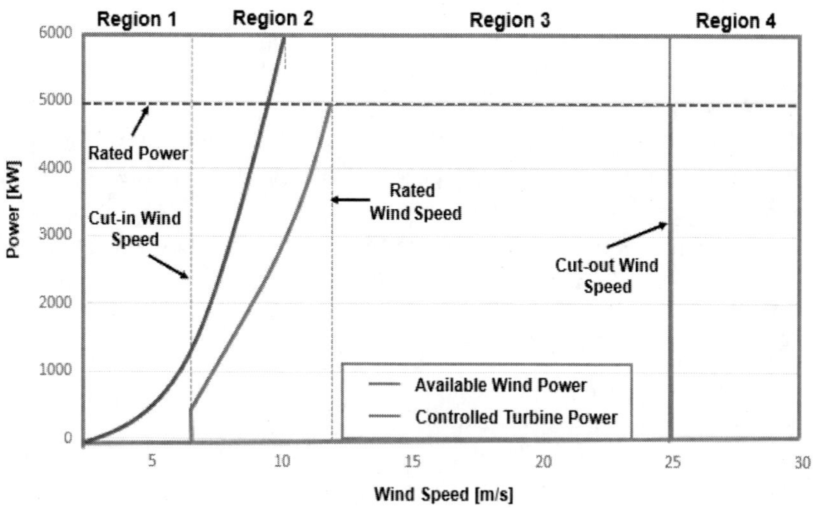

Figure 1. Wind turbine operating regions (Aho et al. 2012).

In Region II, a generator torque controller is typically used to gain the maximum energy possible from the available wind. The controller allows the turbines to have variable rotor speed operation. Therefore, turbines are pushed to be operated at the peak of the Cp-TSR-Pitch surface or shortly Cp surface. Extraction of this surface is defined using the MS Bladed Model in the proceeding pages of this chapter. When the turbine operates with the optimum TSR, it operates with the maximum power coefficient, C_{pmax}. Operating with C_{pmax} the turbine produces the maximum power at any wind speeds at a steady-state. As stated previously, this is obtained by setting the blades at their optimum or fine pitch settings and adjusting the generator torque to change the rotor speed against the changing wind speed. During turbine operation, the Cp surface may vary, and consequently a lower maximum C_p occurs due to residue buildup, blade erosion or blade icing, etc. This phenomenon negatively changes the Cp surface. Thus, an adaptive control technique is required in case of this situation. However, in this chapter, the rotor blades are assumed to be clean, i.e., no icing, or erosion, etc. during the generator torque controller design. In Region III, however, a Proportional and Integral (PI) or Proportional-Integral-Derivative (PID) based blade pitch controller (Johnson et al. 2006) (Merabet, Thongam, and Gu 2011) is most commonly used to produce the rated power, which is the primary purpose in Region III. To ensure that, the generator torque is kept fixed at its rated value and blade pitch angles are adjusted, collectively to regulate the turbine rotor speed.

Generator torque control can be carried out in two different ways. These are torque-mode and speed mode controls. In the standard torque control, the torque demand of the generator is produced from the square of generator or rotor speed times an optimal gain, which is obtained through the use of the turbine and rotor aerodynamic properties. In this chapter, rather than the generator speed, the rotor speed is used for control purposes. A standard nonlinear torque control algorithm for the demanded torque from the generator is used. In the speed control mode, on the other hand, there is usually a PI-based controller to produce the demanded torque from the generator (Lim 2012).

MS Bladed Wind Turbine Simulation Model

The MS (Mustafa Şahin) Bladed Wind Turbine Simulation Model is a nonlinear dynamic turbine model (Sahin 2018). It is developed by the corresponding author of this chapter for the simulations of the Horizontal Axis Wind Turbine System. The MS Bladed Model uses the Blade Element Momentum theory as a core concept of turbine aerodynamics. Also, it includes some other aerodynamic corrections where the BEM theory becomes invalid such as in turbulent wake state, skewed wake rotation, etc. The model employs particular coordinate systems such as the wind-aligned coordinate system, yaw-aligned coordinate system, hub-aligned coordinate systems, etc. It has the property of nacelle yawing and blade pitching, collectively or individually. A desired rotor precone angle and a nacelle tilt angle can be defined in the turbine model. The current version of the MS Bladed Model considers all the turbine parts as rigid structures such as the tower, shafts, blades, etc. Therefore, it represents the turbine by the flowing first-order equation.

$$J_t\dot{\Omega} = \tau_a - \tau_{gen} \quad (1)$$

$$J_t = J_r + N_{gear}^2 J_{gen} \quad (2)$$

where Ω is the rotor speed, τ_a the aerodynamic rotor torque, τ_{gen} the generator electromagnetic torque, the N_{gear} the gearbox ratio, whereas J_t is the turbine total inertia, J_r, turbine rotor inertia, and J_{gen}, electrical generator inertia (Sahin 2018).

When the aerodynamic torque is estimated, and the rotor speed information is taken into account, the turbine power can be obtained by the following equation.

$$P = \tau_{aero}\Omega \quad (3)$$

More information about the theory of the MS Bladed Model is defined in the previous chapter and Ref. (Sahin 2018).

NREL 5 MW Turbine, Cp and Ct Surface Extraction

The NREL 5 MW turbine is an upwind, variable-speed, variable blade-pitch-to-feather controlled turbine with three blades. It has a rated wind speed of 11.4 m/s, cut-in and cut-out wind speeds for the turbine are 3 and 25 m/s, respectively. It has a rotor precone angle of 2.5 degrees and a nacelle tilt angle of 5 degrees (Jonkman et al. 2009).

The chord and twist distributions of the turbine blades are respectively shown in Figure 2-a, and b.

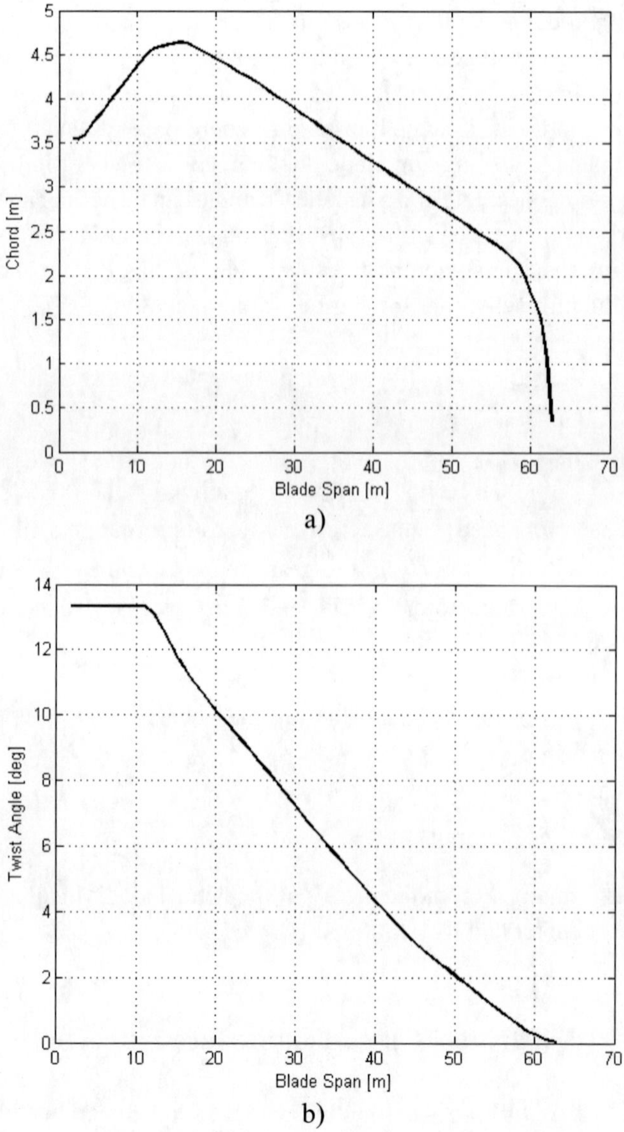

Figure 2. NREL 5 MW wind turbine, a) Chord distribution, b) Twist distribution (Jonkman et al. 2009; Sahin 2018).

The NREL 5 MW turbine rotor blade includes six different airfoils throughout the blade span except for two circular cross-sections at the blade root. As in the FAST simulation Model, the turbine blades are divided into 17 blade element sections in the MS Bladed Model. The NREL 5 MW turbine blade Rsection, their partial span lengths, or DRsection and corresponding airfoils are given in Table 1.

Table 1. NREL 5 MW Turbine Sectional Radius, Span Lengths and Airfoil Distributions (Jonkman et al. 2009)

Section	RSection (m)	DRsection	Airfoils
1	2.8667	2.7333	Cyclinder1
2	5.6000	2.7333	Cyclinder1
3	8.3333	2.7333	Cyclinder2
4	11.7500	4.1000	DU40-A17
5	15.8500	4.1000	DU35-A17
6	19.9500	4.1000	DU35-A17
7	24.0500	4.1000	DU30-A17
8	28.1500	4.1000	DU25-A17
9	32.2500	4.1000	DU25-A17
10	36.3500	4.1000	DU21-A17
11	40.4500	4.1000	DU21-A17
12	44.5500	4.1000	NACA64-A17
13	48.6500	4.1000	NACA64-A17
14	52.7500	4.1000	NACA64-A17
15	56.1667	2.7333	NACA64-A17
16	58.9000	2.7333	NACA64-A17
17	61.6333	2.7333	NACA64-A17

In the MS Bladed Model, the aerodynamic data (Jonkman et al. 2009), which are extrapolated from -180 to +180 degrees, are utilized for these airfoils in Table 1. These data include the effect of the 3D stall delay phenomenon. Some other properties of the NREL 5 MW turbine are found in Ref. (Jonkman et al. 2009).

To design the generator torque controller in this chapter, Cp surface is required. This surface is extracted using a turbine simulation model. Figure 3 shows the Cp surface and its contours obtained using the MS Bladed Model. The optimum blade pitch angle and TSR values, which are used to calculate the torque controller gain, K_c (5) are obtained from the Cp surface. These optimum values are the values that correspond to the maximum Cp value on the surface. Here, the optimum blade pitch angle and TSR are determined as -0.875 degree and 7.5, respectively.

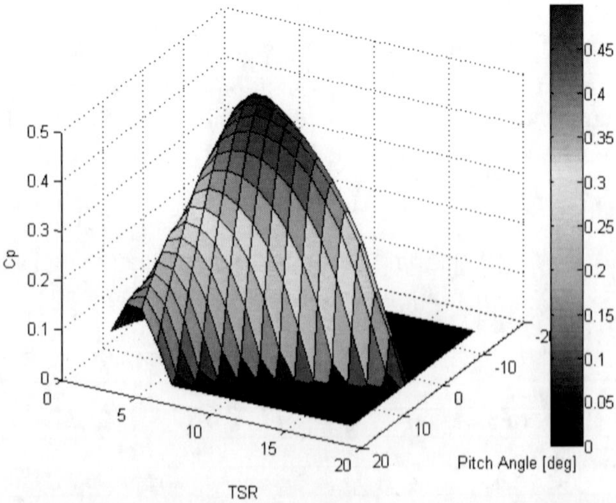

a)

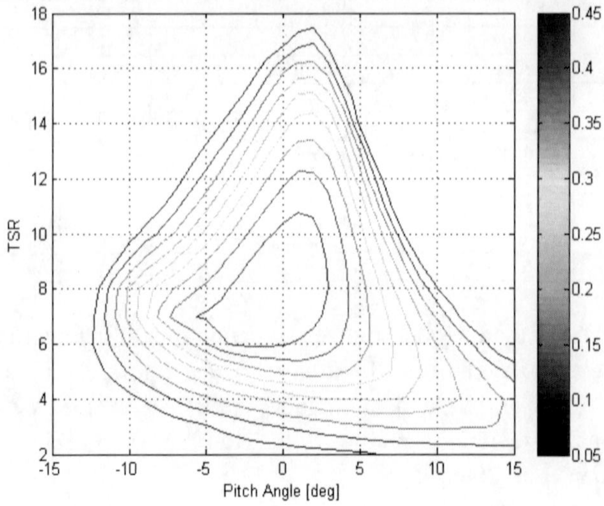

b)

Figure 3. Cp-TSR-Pitch surface, a) Cp surface, b) Cp contours (Sahin 2018).

Figure 4, on the other hand, shows the Ct-TSR-Pitch surface or Ct surface and its contours, which are also extracted using the MS Bladed Model for the NREL 5 MW turbine rotor by similarly operating the rotor at various blade pitch angles and Tip Speed Ratios, or TSRs. Extracting both of these surfaces using the MS Bladed Model takes a moderate amount of time, which depends on the step size/resolution selected for the blade pitch angle and TSR. It is also dependent upon the speed of the utilized computer processor.

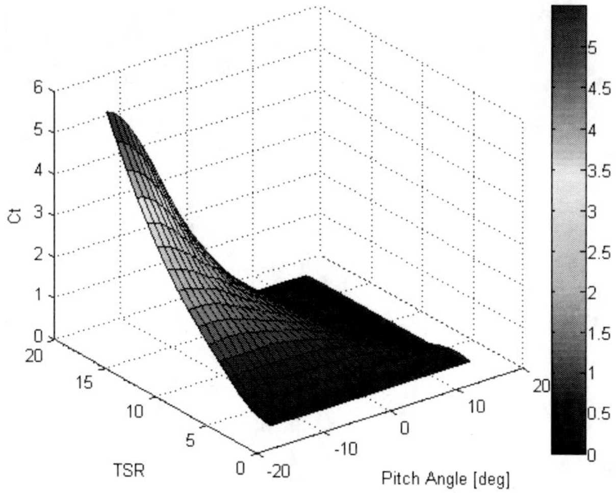

a)

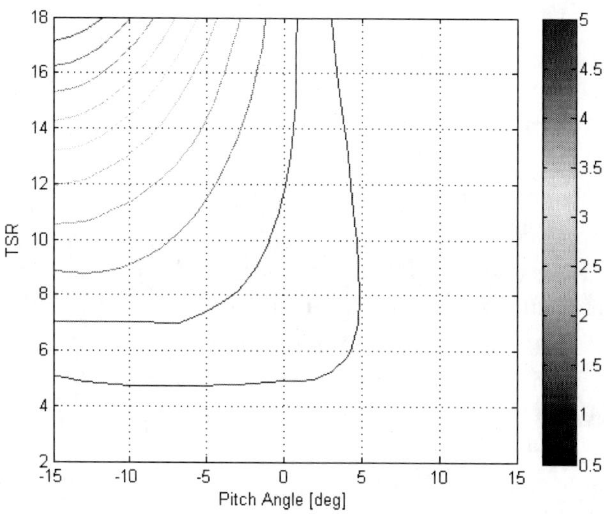

b)

Figure 4. C_T-TSR-Pitch Surface a) C_T surface, b) C_T contours (Sahin 2018).

During Cp and Ct surface extraction, TSR is ranged from 2 to 18 by a step size of 1, while the blade pitch angle is also increased from -15 to 15 degrees by the same step size. As seen from Figure 3 and Figure 4, Cp and Ct surfaces are quite different from each other.

For comparison, Table 2 shows the optimum TSR and blade pitch values determined from the Cp surfaces using the MS Bladed Model and the NREL FAST Model (Jonkman et al. 2009; Sahin 2018). As seen from the table, the values for the maximum Cp and TSR and the blade pitch angle are quite close to each other.

Table 2. Estimation of optimum variables for the NREL 5 MW turbine with different Turbine Simulation Models

Turbine Simulation Model	C_{pmax}	λ_*	β_*
MS Bladed Model	0.4996	7.5	-0.875
NREL FAST Model	0.482	7.55	0

Here, the difference in these values is mainly caused by using different aerodynamic correction models within both models for the turbulent wake state operation. The MS Bladed Model uses Spera's correction (M. O. L. Hansen 2008), while the NREL FAST Simulation Model uses the Buhl correction model (Buhl 2005).

Design of Baseline Turbine Controllers

The properties of the NREL 5 MW turbine are utilized in the MS Bladed Model (Jonkman et al. 2009). A generator torque controller for the below rated region, i.e., Region II, and a collective blade pitch controller for the above rated region, Region III, are designed. In addition, transition region torque controllers, i.e., Region $I^{1/2}$ and Region $II^{1/2}$ are also added to the torque control system. Figure 5 shows the general block diagram for a Wind Turbine, i.e., the NREL 5 MW turbine with the designed generator torque and blade pitch controllers.

As seen in Figure 5, both controllers use the rotor speed information as feedback. Based on this information, generator torque or collective blade pitch controller runs the turbine from cut-in to cut-out wind speed. In this block diagram, the rotor speed reference is given for the blade pitch controller, only.

In addition, the blade pitch and generator torque controller outputs are fed to the blade pitch and torque actuators, respectively.

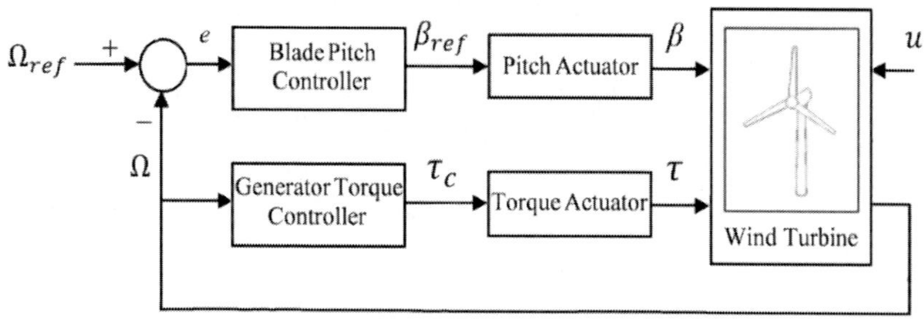

Figure 5. The NREL 5 MW turbine system with the generator torque and blade pitch controllers (Sahin and Yavrucuk 2020).

Generator Torque Controller Design

As declared before, the main goal in Region II, or low to moderate wind speeds, is to obtain the maximum power possible from the turbine rotor. This can be obtained by achieving the maximum aerodynamic efficiency from the rotor, namely operating it at the maximum power coefficient, C_{pmax}. For this purpose, the turbine is required to be operating at a variable rotor speed for the optimum TSR with the blade pitch angles adjusted to the optimum value. Namely, in Region II, all the blade pitch angles are fixed/saturated to the optimum value, and by adjusting/controlling the generator torque, the rotor speed is varied so that the turbine rotor can operate with the optimum TSR value and thereby operating it at the maximum C_p (Johnson et al. 2006; Merabet, Thongam, and Gu 2011; Stol and Fingersh 2004).

In this chapter, a typical nonlinear generator torque controller is designed for Region II of the NREL 5 MW turbine. It is based on the following control law (Johnson et al. 2006; Merabet, Thongam, and Gu 2011; Stol and Fingersh 2004):

$$\tau_c = K_c \Omega^2 \quad (4)$$

where τ_c is the torque controller output, i.e., the demanded electromagnetic torque from the generator, Ω is the rotor speed, while K_c is the torque controller gain and is defined as

$$K_c = \frac{1}{2}\rho\pi(R\cos\Phi)^5\frac{C_{pmax}(\lambda_*,\beta_*)}{\lambda_*^3} \quad (5)$$

where ρ, R, Φ, λ_* and β_* are air density, rotor radius, precone angle, the optimum TSR, and optimum blade pitch angle, respectively. Previously, the optimum TSR, λ_* and optimum blade pitch angles, β_* are respectively obtained as 7.5 and -0.875 degree, whereas C_{pmax} is 0.4996 for the NREL 5 MW turbine (Table 2).

The dynamic turbine model in (1) can also be written considering the τ_{gen} being produced by the torque controller as τ_c. When the equation is solved for $\dot{\Omega}$,

$$\dot{\Omega} = \frac{1}{J_t}(\tau_{aero} - \tau_c) \quad (6)$$

Following References (Pao and Johnson 2009; Johnson et al. 2006; Galvani, Sun, and Turkoglu 2016; Sahin 2018), the theory and idea of how the utilized nonlinear generator torque controller operates the turbine at a variable rotor speed are defined below. Neglecting the yaw effect (Sahin 2018).

$$\lambda = \frac{\Omega R\cos\Phi}{u\cos\theta} \quad (7)$$

The power produced by the turbine is obtained by

$$P = \frac{1}{2}\rho(u\cos\theta)^3\pi(R\cos\Phi)^2C_p \quad (8)$$

Therefore, the aerodynamic rotor torque (τ_{aero}) is obtained using (3), (7), and (8) as

$$\tau_{aero} = \frac{1}{2}\rho\pi(R\cos\Phi)^3(u\cos\theta)^2\frac{C_p}{\lambda} \quad (9)$$

When the aerodynamic torque in equation (9) and generator control torque in equation (4) considering the controller gain, K_c in equation (5) are placed into the equation (6), the following is obtained

$$\dot{\Omega} = \frac{1}{2J_t}\rho\pi\Omega^2(R\cos\Phi)^5\left(\frac{C_p}{\lambda^3} - \frac{C_{pmax}}{\lambda_*^3}\right) \quad (10)$$

In equation (10), as the total moment of turbine inertia, J_t, air density, ρ, rotor radius, R, and rotor speed, Ω, are to be always positive. For the rotor acceleration to be positive or negative depends upon the expressions in parenthesis, or the following conditions (Pao and Johnson 2009).

I. If $\lambda > \lambda_*$, then $C_p \leq C_{pmax}$, the rotor acceleration is negative. Thus, rotor speed decreases till the rotor TSR, λ, becomes equal to the optimum TSR, λ_*. When the equality, $\lambda = \lambda_*$, is satisfied, the rotor is kept turning at the reached rotor speed until a change occurs in wind speed.

II. If $\lambda < \lambda_*$ and $C_p > \frac{C_{pmax}}{\lambda_*^3}\lambda^3$, rotor acceleration becomes positive and thus the rotor starts accelerating until it reaches the optimum TSR. When the equality, $\lambda = \lambda_*$, is satisfied, the rotor turns with constant speed at the reached rotor speed.

The inequality in condition II can be defined by a function, N, which depends on the TSR, λ. (Pao and Johnson 2009).

$$N = \frac{C_{pmax}}{\lambda_*^3}\lambda^3 \qquad (11)$$

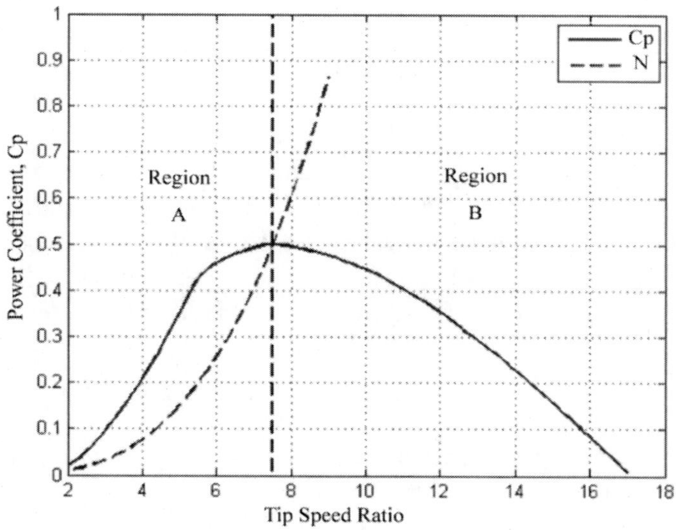

Figure 6. Cp versus λ and N versus λ at the optimum blade pitch angle of -0.875 degree (Sahin 2018).

Also, Cp versus TSR and N versus TSR for the NREL 5 MW turbine can be obtained as in Figure 6 using the MS Bladed Model. Note that the Cp versus λ in Figure 6 is a part of the Cp surface in Figure 3.

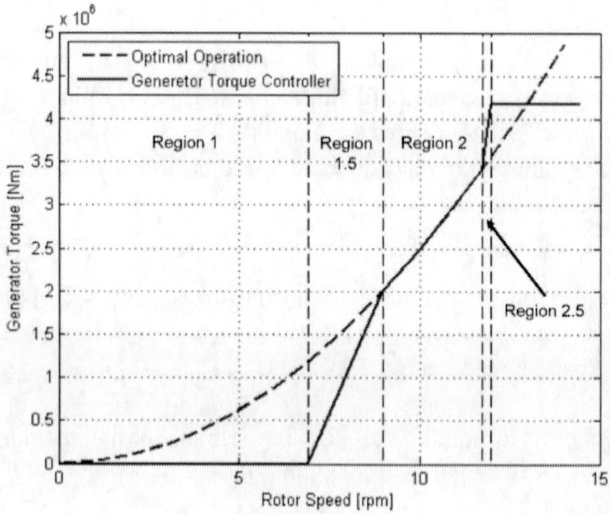

Figure 7. Generator torque controller output, generator torque on LSS versus rotor speed (Sahin 2018).

In Figure 6, the crossing point of Cp and N curves is the equilibrium operating point for the 5 MW turbine. When the turbine operates with condition I, then the turbine operates in Region B, otherwise, the operation falls in Region A. Also, to switch from one region to another, i.e., from Region I to II or from Region II to III, transition region torque controllers, Region $I^{1/2}$ and Region $II^{1/2}$, are designed. These are linear torque lines and switching between the controllers is realized dynamically depending on the rotor/generator speed (Petrović and Bottasso 2017; Jonkman et al. 2009). Design of both transition torque controllers are the same and is defined as

$$\tau_c = m(\Omega - \Omega_l) + \tau_l \qquad (12)$$

where m is the slope of the corresponding transition line in the related transition region (Figure 7) and is obtained as

$$m = \frac{\tau_u - \tau_l}{\Omega_u - \Omega_l} \qquad (13)$$

where τ_u, Ω_u are the upper limits for the generator torque and rotor speed, while τ_l, Ω_l are the lower limits for the same parameters in Region $I^{1/2}$ and Region $II^{1/2}$.

In this chapter, since the rotor speed (LSS) is utilized as feedback, the above upper and lower limits are re-obtained from those of the generator considering the gearbox ratio (Jonkman et al. 2009). Therefore, in Figure 7, the generator torque is the torque on the LSS side of the gearbox.

In Region III, the generator torque is kept at its rated value. Figure 7 shows the designed generator torque controller output based on the rotor speed. The above upper and lower limits for the transition regions are also seen in Figure 7.

Since, in the below rated region, a different amount of generator torque is demanded from the generator against changing rotor speed due to the variations in wind speed, a region selector, i.e., IF condition, is used in the torque controller to activate which part of the torque controller, Region $I^{1/2}$, Region II or Region $II^{1/2}$ to produce the output torque or control torque, τ_c, based on the rotor speed. The torque actuator has usually a fast dynamics and therefore is neglected.

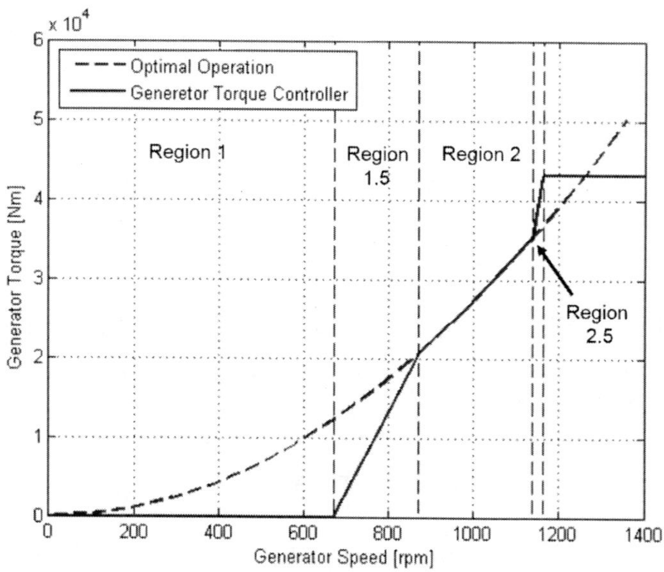

Figure 8. Generator torque controller output, generator torque on HSS versus generator speed (Sahin 2018).

If the generator speed were used as feedback to the controller, the generator torque controller design would be as in Figure 8.

Blade Pitch Controller Design

In Region III, a gain scheduled PI-based collective blade pitch controller is designed. The nonlinear NREL 5 MW turbine model in (1) is linearized around two different equilibrium operating points through the perturbation technique. In Region III, an equilibrium operating point is a point at which wind speed and blade pitch angle, the aerodynamic torque reaches the rated generator torque at the rated rotor speed. The generator torque is assumed to be constant in Region III, therefore, the nonlinearity in the HAWT system is caused by the aerodynamic torque, only (Jonkman et al. 2009). Therefore, the rotor aerodynamic torque in equation (1) is a continuous function of three variables: rotor speed, Ω, blade pitch angle, β, and lastly the wind speed, u. Thus, following Ref. (A. Wright 2004; Sahin and Yavrucuk 2019), the aerodynamic torque may be expanded by a Taylor series as,

$$\tau_{aero}(u,\Omega,\beta) = \tau_{aero}(u_o,\Omega_o,\beta_o) + \frac{\partial\tau_{aero}}{\partial U}(u-u_o) + \frac{\partial\tau_{aero}}{\partial\Omega}(\Omega-\Omega_o) + \frac{\partial\tau_{aero}}{\partial\beta}(\beta-\beta_o) + HOTs \qquad (14)$$

where u_o, Ω_o, β_o are respectively the values of wind speed, shaft speed, and blade pitch angle at an equilibrium operating point, while $u - u_o = \Delta u, \Omega - \Omega_o = \Delta\Omega$ and $\beta - \beta_o = \Delta\beta$ are the perturbations from the operating point. For linear controller designs, Higher Order Terms (HOTs) in (14) are ignored. If the series expansion is applied to the whole HAWT system in (1) considering the generator torque on the LSS of the gearbox, τ_{gen}, the following is obtained.

$$J_t(\Omega \dot{-} \Omega_e) = \tau_{aero}(\Omega_o,\beta_o,u_o) + \gamma\Delta\Omega + +\eta\Delta\beta + \mu\Delta u - \tau_{gen} \qquad (15)$$

For Region III, the rated generator torque on the LSS of the gearbox is determined to be 4180074.35 Nm (Jonkman et al. 2009). Also, there is no rotor acceleration because the aerodynamic and generator torques cancel each other at equilibrium turbine operation. Therefore, for any operation at steady-state,

$$J_t\dot{\Omega} = \gamma\Delta\Omega + \eta\Delta\beta + \mu\Delta u \qquad (16)$$

where, γ, η, μ represent the partial derivatives of $\frac{\partial\tau_{aero}}{\partial\Omega}$, $\frac{\partial\tau_{aero}}{\partial\beta}$ and $\frac{\partial\tau_{aero}}{\partial u}$, respectively. Further,

$$\dot{\Omega} = \frac{\gamma}{J_t}\Delta\Omega + \frac{\eta}{J_t}\Delta\beta + \frac{\mu}{J_t}\Delta u \qquad (17)$$

Defining $\frac{\gamma}{J_t} = A$, $\frac{\eta}{J_t} = B$ and $\frac{\mu}{J_t} = B_d$, the open-loop turbine system becomes

$$\dot{\Omega} = A\Delta\Omega + B\Delta\beta + B_d\Delta u \qquad (18)$$

where A is the turbine system gain, B is the input gain, and finally B_d is the disturbance gain. They are the ratios of the partial derivative of aerodynamic torque with respect to rotor speed, blade pitch angle, and wind speed to the HAWT total inertia, respectively. Following the Ref. (A. D. Wright and Fingersh 2008), the perturbation of pitch angle, $\Delta\beta$, is related directly to that of rotor speed, $\Delta\Omega$,

$$\Delta\beta(t) = K_p\Delta\Omega(t) + K_i\int\Delta\,\Omega(t)dt \qquad (19)$$

where K_p and K_i represent the proportional and integral gains, respectively. Thus, the controller output is as follows, $\beta = \Delta\beta + \beta_o$. Putting equation (19) into equation (18) constructs the closed-loop turbine system with PI strategy as in equation (20).

$$\dot{\Omega} = A\Delta\Omega + B\left(K_p\Delta\Omega(t) + K_i\int\Delta\,\Omega(t)dt\right) + B_d\Delta u \qquad (20)$$

Taking the Laplace Transform of equation (20) as follows,

$$s\Omega(s) = A\Delta\Omega(s) + B\left(K_p\Delta\Omega(s) + \frac{K_i}{s}\Delta\Omega(s)\right) + B_d\Delta u(s) \qquad (21)$$

and then carrying out some algebraic manipulations, the closed-loop transfer function between the rotor speed, $\Delta\Omega(s)$ and the wind, $\Delta u(s)$ is obtained as,

$$G_{CL}(s) = \frac{\Delta\Omega(s)}{\Delta u(s)} = \frac{B_d s}{s^2+(-A-BK_p)s+(-BK_i)} \quad (22)$$

The denominator of the transfer function or the characteristic equation (22) of the closed-loop system gives information about the system stability considering the controller gains, K_p, and K_i. Therefore, by designing K_p, and K_i, the desired performance is obtained from the closed-loop turbine system. However, proper gains should be chosen according to the design requirements such as the natural frequency, w_n and damping ratio, ζ. For wind turbines, the natural frequency is suggested to be 0.6-07 in Ref. (M. H. Hansen et al. 2005), while the value of the damping ratio may change from turbine to turbine. For instance, after several trials, Wright and Fingersh (A. D. Wright and Fingersh 2008) have chosen a damping ratio of 1 for the NREL CART turbine during the design of the blade pitch controller. Therefore, a similar approach in Ref. (A. D. Wright and Fingersh 2008) is used and after a thoughtful investigation, the damping value of 0.8 is decided as the best value (Sahin 2018; Sahin and Yavrucuk 2019).

Using the central difference theorem based on the equilibrium wake assumption (Sahin and Yavrucuk 2019), the linearization of the NREL 5 MW turbine is carried out around an equilibrium point to find the values of partial derivatives ɣ and η to obtain A and B gains in (22). Considering the characteristic equation as a second-order system, K_p and K_i gains are calculated by equations (23) and (24), respectively in terms of natural frequency and damping ratio.

$$K_p = \frac{-2w_n\zeta}{B} - \frac{A}{B} \quad (23)$$

$$K_i = \frac{-w_n^2}{B} \quad (24)$$

A gain scheduling method is applied as follows due to the improper performance of one linear controller in Region III, which is also declared in Ref. (A. D. Wright and Fingersh 2008).

Firstly, an equilibrium operating point close to the entry of Region II into Region III, the OP1 in Table 3, is selected. Afterward, through linearization, the system gain, A, input gain, B, and disturbance gain, B_d for the NREL 5 MW turbine are obtained respectively as -0.0554, -0.2658, and 0.0227. Also, when the desired damping ratio, ζ, and natural frequency, w_n are used respectively as 0.8 and 0.6, the proportional gain, K_p and integral gain, K_i are obtained through equations (23) and (24), respectively as 3.4033, and 1.3544. Here, the design target is to get a rotor speed response with a shorter decay time, minimum overshoot, and without oscillations (A. D. Wright and Fingersh 2008). For the gain scheduling purpose, a gain correction factor formula $GK(\beta)$ in (25) is used (Jonkman et al. 2009; A. D. Wright and Fingersh 2008). By simply multiplying the calculated controller gains by the correction factor, GK, the controller gives a superior performance at any operating point in the above rated region, i.e., Region III.

$$GK(\beta) = \frac{1}{\left(1+\frac{\beta}{\beta_K}\right)} \qquad (25)$$

Table 3. Selected operating points for controller design (Sahin 2018)

Operating Points	Wind Speed (m/s), u_o	Rotor Speed (rpm), Ω_o	Pitch Angle (deg), β_o	Rotor Torque (Nm),τ_o
OP1	11.5	12.1	2.2792	4180074.35
OP2	12.6607	12.1	5.9676	4180074.35

where β is the turbine blade pitch angle required for the turbine rotor to produce the rated torque at any wind speed while operating at the rated rotor speed. According to Wright and Fingers (A. D. Wright and Fingersh 2008), β_K is the blade pitch angle at which the input gain, B, calculated at an equilibrium operating point close to the border of Region II into III has doubled in its value at another equilibrium operating point further in Region III (A. D. Wright and Fingersh 2008). By the linearization of the NREL 5 MW turbine, the OP2 at which the input gain, B has doubled in its value is estimated in Region III. At OP2, β_K is obtained as 5.9676 degrees and is utilized in the calculation of $GK(\beta)$. Here, the blade pitch, β in GK formula (25) is obtained by the interpolation of pitch angles with respect to wind speeds. Afterward, the above proportional, K_p and integral, K_i gains are multiplied by the gain correction factor, GK. Here, the pitch controller is based on the pitch to feather control method. As also seen in the block diagram (Figure 5), e is the error between rotor speed reference, Ω_{ref} and the actual rotor speed measurement, Ω. In the above design, the controller output is β, which is the blade pitch angle input to the turbine. Since the MS Bladed Model does not include a pitch actuator dynamic effect, a first-order transfer function (26) with a time constant of 0.2 (A. Wright 2004) is added between the controller output and blade pitch input. A rate limiter of 8 deg/s, and a saturation limit of -0.875 and +90 degrees are also added to the pith control system.

$$\frac{\beta}{\beta_{ref}} = \frac{1}{0.2s+1} \quad (26)$$

Next, the blade pitch controller output is renamed with β_{ref} and therefore the actuator output becomes β, the blade pitch angle input to the MS Bladed Model. Also, an anti-windup is added to the pitch control system as in Ref. (A. D. Wright and Fingersh 2008) to avoid undesired rotor speed response under a strong wind gust that takes the turbine from a Region II operation to a Region III operation. Without the anti-windup, the rotor speed response reaches large values under a large step increasing wind speed. The turbine blade pitch response sharply reaches a large value and then gets its steady-state within a very short time. This is not feasible and is probably dangerous for a turbine blade in the length of 63-meters. More information about the anti-windup and simulations can be found in Ref. (Sahin 2018).

Simulation of the Controlled NREL 5 MW Turbine

In the previous section, baseline generator torque and collective blade pitch controllers are designed. In this section, their simulations are carried out using the MS Bladed Model for the NREL 5 MW turbine (Jonkman et al. 2009).

Dynamic Operation of the Controlled NREL 5 MW Turbine

Here, a step wind speed increasing at every 25s, and starting from a value of 7 m/s reaching to 14 m/s (Figure 9-a) is used. Therefore, this wind speed drives the controlled 5 MW turbine throughout all its operating regions such as Region I$^{1/2}$, Region II, Region II$^{1/2}$, and Region III. When the wind speed (Figure 9-a) is applied during a steady-state turbine operation, turbine responses change, drastically at around 150s, indicating a transition process from one controller to another, i.e., from the generator torque controller to the blade pitch controller.

The NREL 5 MW turbine is controlled by the generator torque controller until around 150s, while in the rest of the simulation, it is controlled by the blade pitch controller. At 150s, the wind speed increases from 11 m/s to 12 m/s surpassing the rated wind speed of 11.4 m/s, sharply.

Approximately, below 30s corresponds to the Region $I^{1/2}$ operation, between 30 and 120s is the Region II operation, while 120 and 150s is a transition region (Region $II^{1/2}$) operation. Below 150s, the generator torque controller is active. Above 150s is a Region III operation and therefore the blade pitch controller manages the 5 MW turbine operation.

In Region II as well as the transition Region $I^{1/2}$ and $II^{1/2}$, the blade pitch angle is saturated by the blade pitch controller and is kept constant at its optimum value of -0.875 degree (Figure 9-b). Whenever the wind speed increases (Figure 9-a), the aerodynamic torque of the turbine rotor increases (Figure 9-c), while the turbine TSR decreases (Figure 10-e). Turbine TSR decreases since it is inversely proportional to the wind speed (equation 7).

Therefore, the turbine operating point moves to another operating point with a lower Cp value than the Cpmax of 0.4996. Then, turbine operation falls into Region A (Figure 6). Therefore, the rotor speed is expected to accelerate for the turbine rotor to reach the optimum TSR of 7.5. Thus, the generator torque controller adjusts the demanded torque from the generator and increases the electromagnetic torque and a new equilibrium operating point is obtained (Figure 9-c) and the turbine operates with the optimum TSR (Figure 10-e), and the Cpmax (Figure 10-f) in Region II. In Region $I^{1/2}$, the turbine operates with a TSR higher than the optimum value and thereby with a lower Cp value than the Cpmax. In Region $II^{1/2}$, the turbine switches from Region II to Region III. In this region, turbine TSR starts decreasing, and therefore it operates with a lower Cp value.

In Region III, the blade pitch controller takes the turbine control. The controller increases the turbine blade pitch angles, collectively (Figure 9-b) against the step increasing wind speed (Figure 9-a), and regulates the rotor speed at its rated value of 12.1 rpm (Figure 9-d). When the wind speed increases, the turbine TSR (Figure 10-e), and Cp (Figure 10-f) get lower and lower. Figure 10-g and h show the changes in turbine power and thrust force, respectively. Below the rated wind speed, the turbine produces a power lesser than the 5 MW rated power. In the above rated region, the rated power is achieved. When the wind speed increases, the thrust of the turbine increases in the below rated region, but decreases in the above rated region.

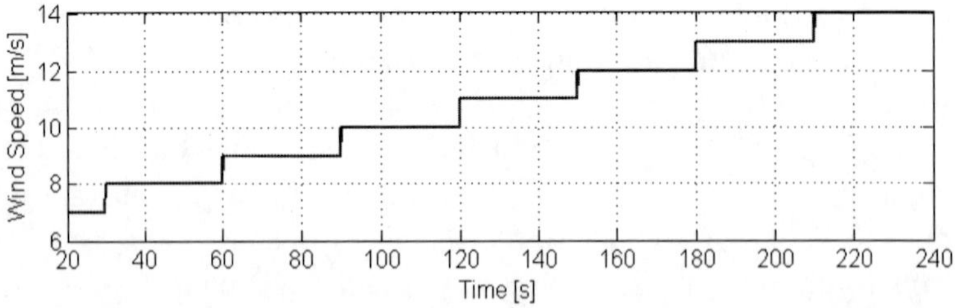

a)

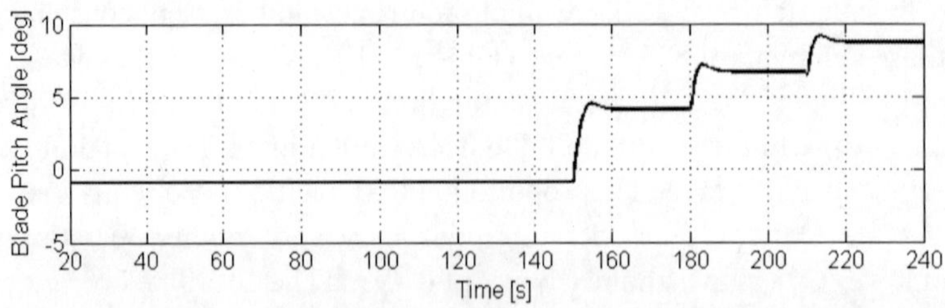

b)

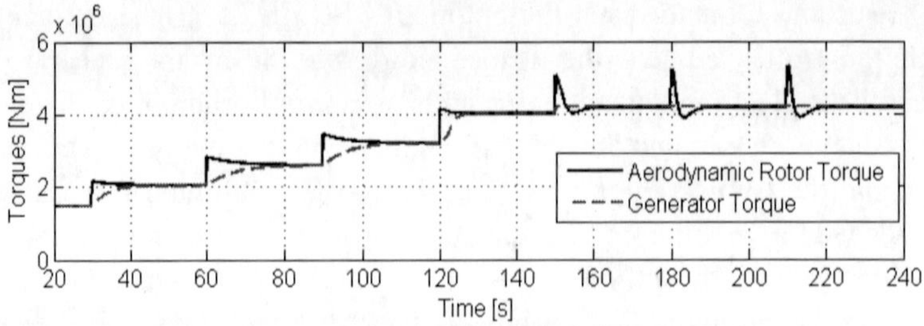

c)

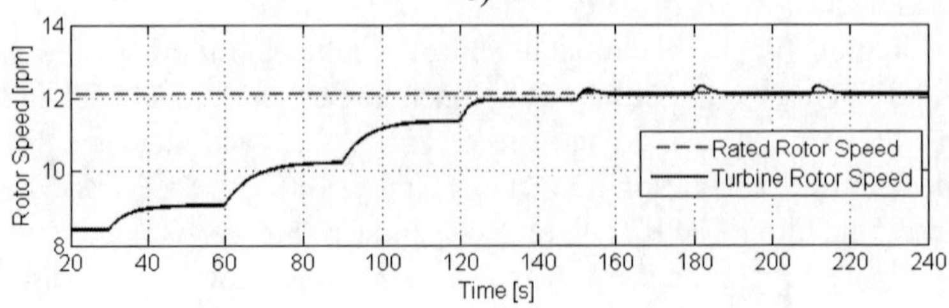

d)

Figure 9. Controlled NREL 5 MW turbine simulations, a) Wind speed, b) Blade pitch angle, c) Torques on LSS d) Rotor speed, e) Tip Speed Ratio, f) Power coefficient, g) Power, h) Thrust force.

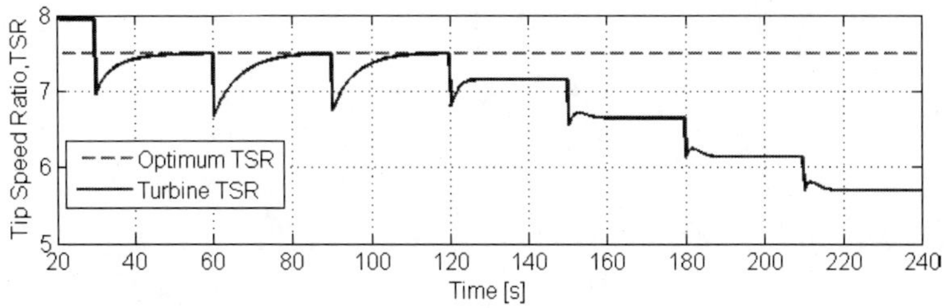

e)

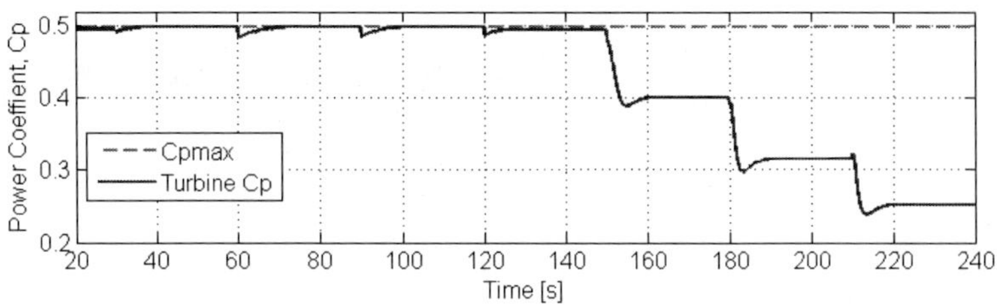

f)

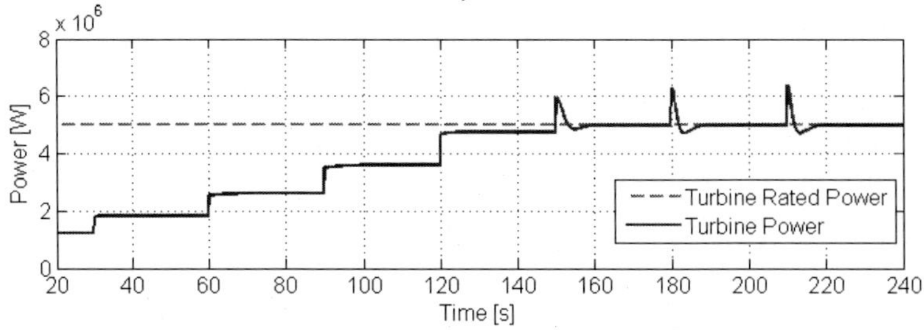

g)

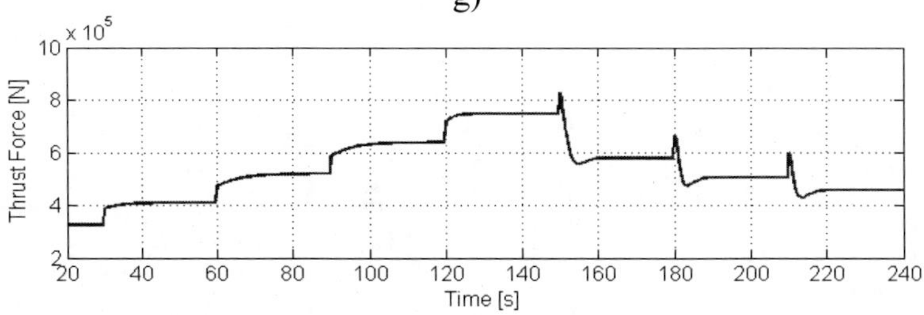

h)

Figure 10. Controlled 5 MW turbine simulations (Continue), a) Wind speed, b) Blade pitch angle, c) Torques on LSS d) Rotor speed, e) Tip Speed Ratio, f) Power coefficient, g) Power, h) Thrust force.

Steady-State Operation of the Controlled NREL 5 MW Turbine

Here, steady-state responses of the controlled NREL 5 MW turbine are examined. Different levels of constant wind speeds starting from the cut-in to cut-out wind speed are applied to the controlled NREL 5 MW turbine with generator torque and collective blade pitch controller. The changes in rotor speed, blade pitch, power, etc. are obtained from simulations as follows.

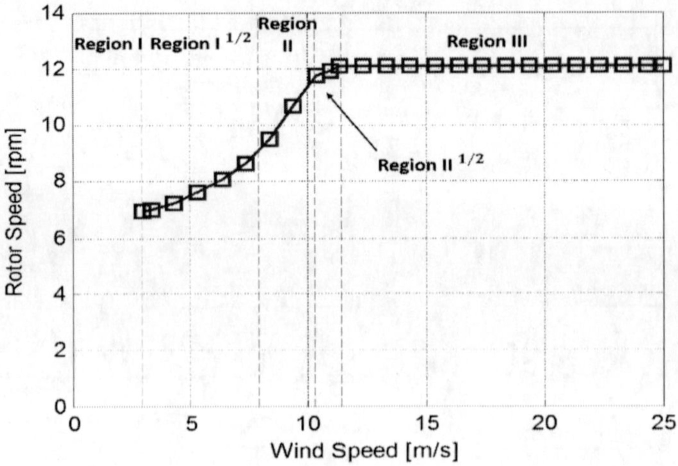

a)

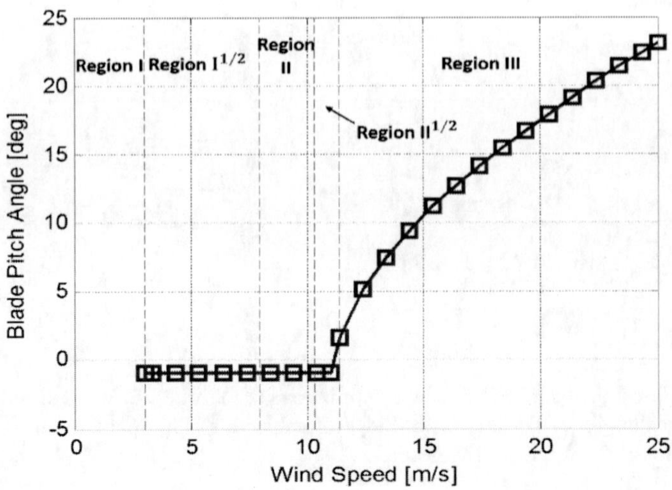

b)

Figure 11. Rotor speed and blade pitch angle versus wind speed.

As seen in Figure 11-a, in the below rated region, (Region $I^{1/2}$, Region II, and Region $II^{1/2}$) the rotor speed is varied with respect to the wind speed, i.e., variable speed operation. When the wind speed is increased, the rotor speed is increased. It is desired not to increase the rotor speed further when the wind speed reaches around rated wind speed. This is why a torque speed ramp is used in Region $II^{1/2}$. In the above rated region, or Region III, the rotor speed is regulated at its rated value of 12.1 rpm and does not change with the change in wind speed. This is because the blade pitch controller regulating the rotor speed by adjusting the turbine blade pitch angles, collectively. In the below rated region, the blade pitch angle is adjusted to its optimum value of -0.875, but it increases when the wind speed increases in the above rated wind speed (Figure 11-b). The rotor speed response is shaped by both generator torque and collective blade pitch controllers from cut-in to cut-out wind speed.

Figure 12-a and b show the controlled power and thrust curves versus wind speed. In the below rated region, the turbine produces power lesser than the rated value. Therefore, below rated region is also called the partial load region, whereas the above rated region is called the full load region since the turbine produces the rated power. In below rated region, when the wind speed increases the thrust force increases while in the above rated region, the thrust force starts decreasing with the increase in wind speed. The thrust force reaches its maximum value around the rated wind speed of 11.4 m/s. This behaviour is only observed for the pitch to feather controlled wind turbines. This thrust peak is so critical that the wind power industry can utilize important strategies in order to get rid of this undesired excessive thrust forces. The industry utilizes algorithms so referred to as peak shaving or thrust clipping algorithm, which is activated around the rated wind speed.

The TSR value decreases when the wind speed increases, except for Region II, where the optimum TSR operation is achieved by the generator torque controller for the maximum aerodynamic efficiency (Figure 13-a). The Cp value starts increasing at Region $I^{1/2}$, and reaches its maximum value in Region II. It starts decreasing in Region $II^{1/2}$ and effectively reduces when the wind speed increases more into Region III up to the cut-out wind speed of 25 m/s (Figure 13-b).

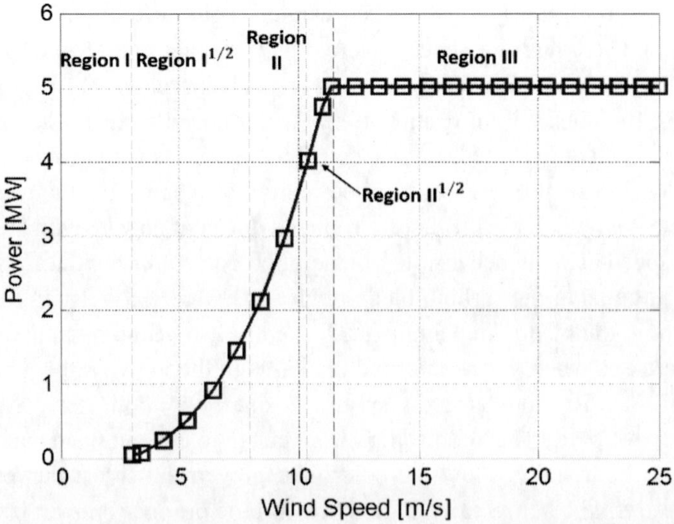

a)

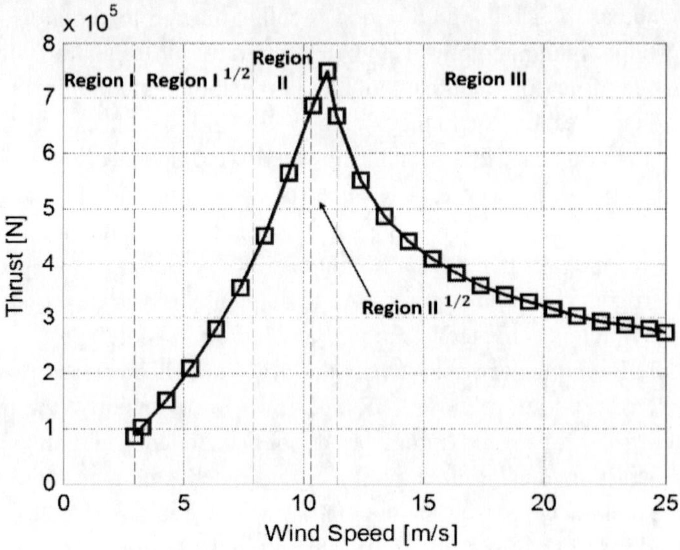

b)

Figure 12. Power and thrust versus wind speed.

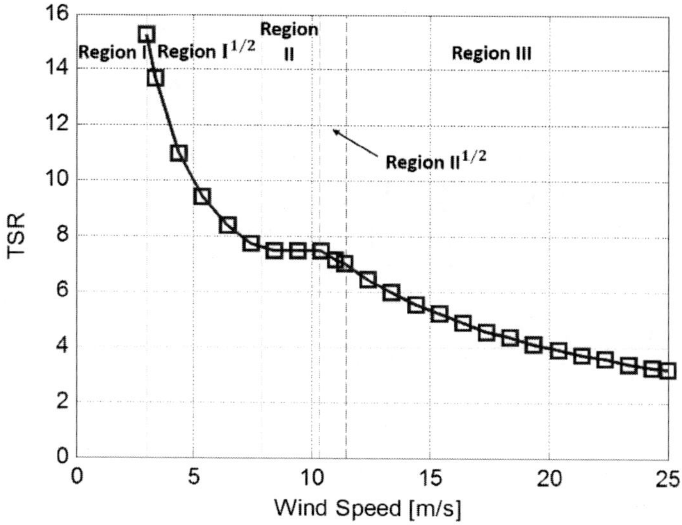

a)

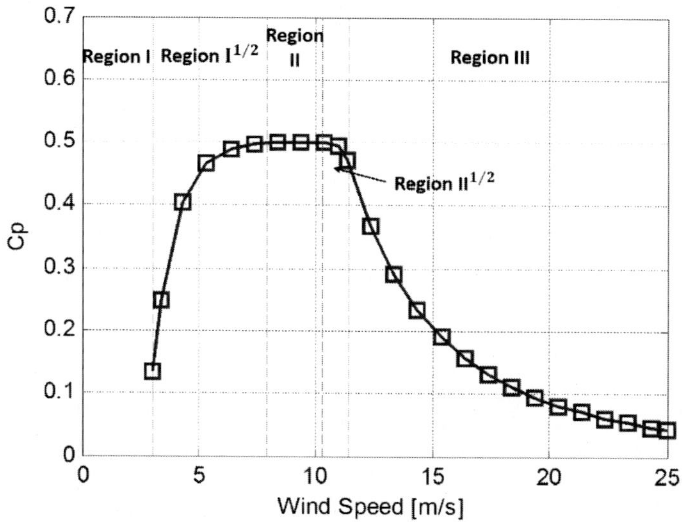

b)

Figure 13. Tip Speed Ratio and power coefficient versus wind speed.

Conclusion

In this chapter, the theory and design of the baseline controllers such as generator torque, and blade pith controllers are investigated in detail. The design of these controllers is realized using the MS Bladed Model for the NREL 5 MW turbine. Also, HAWT systems, control and operating regions for the variable speed variable pitch wind turbines are discussed. The MS Bladed Model is briefly defined. The properties of the NREL 5 MW turbine are also given shortly. Extracting Cp and Ct surfaces are covered. After the design of both controllers, the controlled NREL 5 MW turbine is simulated under step increasing wind speeds ranging from an operating point in Region II to another operating point in Region III, i.e., between wind speeds of 7 and 14 m/s. Dynamic simulation results are obtained in terms of various turbine parameters such as blade pitch angle, generator and rotor torques, rotor speed, TSR, power coefficient, power and thrust forces. Lastly, the steady-state turbine responses of the 5 MW turbine are obtained from cut-in wind speed of 3 m/s to cut-out wind speed of 25 m/s in terms of rotor speed, blade pitch, power, and thrust force, TSR, and Cp.

Acknowledgment

The corresponding author is very grateful to Dr. Jason Jonkman from the US National Renewable Energy Laboratory (NREL) for his valuable guidance, comments, and sharing experiences during the design of baseline turbine controllers for the NREL 5 MW turbine.

References

Aho, J., A. Buckspan, J. Laks, P. Fleming, Yunho Jeong, F. Dunne, M. Churchfield, L. Pao, and K. Johnson. (2012). "A Tutorial of Wind Turbine Control for Supporting Grid Frequency through Active Power Control." *2012 American Control Conference (ACC)*, no. March: 3120–31. https://doi.org/10.1109/ACC.2012.6315180.

Boukhezzar, B., L. Lupu, H. Siguerdidjane, and M. Hand. (2007). "Multivariable Control Strategy for Variable Speed, Variable Pitch Wind Turbines." *Renewable Energy* 32 (8): 1273–87. https://doi.org/10.1016/j.renene.2006.06.010.

Buhl, M. L. (2005). "A New Empirical Relationship between Thrust Coefficient and Induction Factor for the Turbulent Windmill State A New Empirical Relationship between Thrust Coefficient and Induction Factor for the Turbulent Windmill State." *Technical Report NREL/TP-500-36834*.

Fischer, Boris, and Martin Shan. (2013). "A Survey on Control Methods for the Mitigation of Tower Loads."

Galvani, P. A., F. Sun, and K. Turkoglu. (2016). "Aerodynamic Modelling of the NREL 5-MW Wind Turbine for Nonlinear Control System Design: A Case Study Based on Real-Time Nonlinear Receding Horizon Control." *Aerospace 2016*. https://doi.org/10.3390/aerospace3030027.

Hai, J., W. Sun, G. Hu, and A. Wu. (2012). "The Study on Blade Pitch Control of Wind Power with Gain Scheduling." *Key Engineering Materials* 522: 364–68.

Hansen, M. O. L. (2008). *Aerodynamics of Wind Turbines*. 2th Edit. Earthscan.

Hansen, Morten H, Anca Hansen, Torben J Larsen, Stig Øye, Poul Sørensen, and Peter Fuglsang. (2005). "Control Design for a Pitch-Regulated, Variable Speed Wind Turbine, Risø-R-1500(EN),Risø National Laboratory." Roskilde, Denmark.

Hemami, Ahmed. (2012). *Wind Turbine Technology*. Clifton Park, NY 12065-2919 USA: Cengage Learning.

Johnson, K. E., L. Y. Pao, M. J. Balas, and L. J. Fingersh. (2006). "Control of Variable Speed Wind Turbines-Standard and Adaptive Techniques for Maximizing Energy Capture." *IEEE Control Systems Magazine* 26 (3): 70–81.

Jonkman, J., S. Butterfield, W. Musial, and G. Scott. (2009). "Definition of a 5-MW Reference Wind Turbine for Offshore System Development." *Technical Report NREL/TP-500-38060*.

Letcher, Trevor M. (2017). *Wind Energy Engineering-A Handbook for Onshore and Offshore Wind Turbines*. Edited by Trevor M. Letcher. Elsevier Inc.

Lim, C. (2012). "Torque Control Methods for Wind Turbines and Their Responses." In *12th International Conference on Control, Automation and Systems*, 1788–90. ICC, Jeju Island, Korea.

Merabet, A., J. Thongam, and J. Gu. (2011). "Torque and Pitch Angle Control for Variable Speed Wind Turbines in All Operating Regimes." In *10th International Conference on Environment and Electrical Engineering (EEEIC), 2011*, 1:1–5. Roma: IEEE.

Oudah, Ali, II Mohd, and A Hameed. (2014). "Modern Control Techniques in Wind Turbine." *International Journal of Hybrid Information Technology* 7 (4): 101–22. http://www.sersc.org/journals/IJHIT/vol7_no4_2014/10.pdf.

Pao, L. Y., and K. E. Johnson. (2009). "A Tutorial on the Dynamics and Control of Wind Turbines and Wind Farms." In *2009 American Control Conference*, 2076–89. Hyatt Regency Riverfront, St. Louis, MO, USA. https://doi.org/10.1109/ACC.2009.5160195.

Petrović, Vlaho, and Carlo L. Bottasso. (2017). "Wind Turbine Envelope Protection Control over the Full Wind Speed Range." *Renewable Energy* 111: 836–48. https://doi.org/10.1016/j.renene.2017.04.021.

Sahin, M. (2018). "Dynamic Modeling, Control and Adaptive Envelope Protection System for Horizontal Axis Wind Turbines, PhD Thesis, Department of Aerospace Engineering, METU, Ankara." Middle East Technical University.

Sahin, M., and I. Yavrucuk. (2019). "Performance Comparison of Two Turbine Blade Pitch Controller Design Methods Based on Equilibrium and Frozen Wake Assumptions." In *Ankara International Aerospace Conference, AIAC 2019*, 1–16. Ankara, Turkey.

Sahin, M., and İ. Yavrucuk. (2020). "Adaptive Envelope Protection Control for the below and above Rated Regions of Wind Turbines." *International Journal of Energy and Power Engineering* 14 (10): 275–83.

Stol, K. A., and L. J . Fingersh. (2004). "Wind Turbine Field Testing of State-Space Control Designs." *Technical Report NREL/SR-500-35061*, 2004.

Wright, A. D., and L. J. Fingersh. (2008). "Advanced Control Design for Wind Turbines Part I : Control Design, Implementation, and Initial Tests." *Technical Report of National Renewable Energy Laboratory*, no. March: 98.

Wright, Alan. (2004). "Modern Control Design for Flexible Wind Turbines." *Technical Report of National Renewable Energy Laboratory, NREL*, no. July: 233.

Chapter 3

Energy Potential Wind Facility Calculation from Field Measurements and Technical Parameters of the Wind Farm Using Weibull Distribution Methodology

Jesús Ordoñez-Guerrero[*]
Department of Electrical, Systems and Automation Engineering,
University of León, León, Spain

Abstract

This proposed chapter develops the methodology for calculating the annual wind or energy potential production of an installation using the Weibull distribution method, based on wind measurements made at the site object of the project and technical parameters of the power plant, mainly of the wind turbines that will be installed.

The method begins with the application of the Weibull distribution law to the data collected in the area, which is, performing the statistical treatment of the wind data. Starting from this base, the main values that describe this function will be calculated, the scale factor (c) and the shape factor (k). Next, the values of annual mean velocity, median, variance, standard deviation, annual mean of the cube of the velocities and effective velocity of the measurement point will be obtained. The end of this process is the calculation of the wind potential and wind energy density available both at the measurement point and at the wind turbine, since it is the swept area that determines the wind power and energy based on the machine to install. After this initial process, the distribution of velocities and accumulated wind velocity is calculated based on the

[*]Corresponding Author's Email: jesus.ordonez.guerrero@gmail.com.

In: Wind Turbines
Editors: David Borge-Diez and Daniel Icaza
ISBN: 978-1-68507-974-1
© 2022 Nova Science Publishers, Inc.

annual data collection process that gives us the final calculations regarding the energy potential of the wind and wind turbine. At this point, it must be taken into account that the hub height of the possible wind turbine is different from the height of the measurement point. Hellman's law and the Justus and Mikjail equations will be applied to calculate the mean velocity, form factor and scale of the wind due to the change in height. Every parameter described above will be recalculated below, until the new and accumulated velocity distribution is obtained.

The next step is the calculation of the real wind potential based on the hub height, the wind turbine itself and the statistical results obtained. Other parameters such as the power curve of the wind turbine model, estimated losses (electrical, wind farm trails, availability and others), again the swept area and the total number of wind turbines will be used. Thus, the distribution of the wind, the electrical power of the wind turbine and the wind energy density of the wind turbine will be obtained to finish with the calculation of the annual wind turbine production for each type of wind. The following final parameters will be calculat-ed:

- Gross and net wind density and specific production of each wind turbine.
- Annual global electrical performance.
- Load or capacity factor.
- Gross and net production of each wind turbine and of the complete wind farm.
- Equivalent hours and operating factor.

This proposed chapter will include in detail the tables, figures and equations that complement the the calculation methodology set out in this summary document. This is a very useful advance for previous studies of wind resource in an area that additionally calculates its potential including fundamental parameters such as the wind turbine. On the other hand, in the case of adding measurement points in another area within the wind farm site or additional annual periods of time, it is possible to incorporate them into the methodology for obtaining new annual production data in different points or consecu-tive years of measurement. In fact, this would be the most logical process for calculating more real production values. In summary, the main novelty of this chapter is to provide an approximation principle for the correct assessment of the wind potential area based on the wind resource but also on the parameters of the wind turbine to be exploited in the area. On the other hand, the fact that the methodology is able to add more measurements and analysis points results in greater robustness as well as greater precision.

Keywords: atmospheric wind, wind data treatment, wind velocity distribution, weibull distribution, wind power resource, wind turbine parameters

Introduction

The atmosphere is a layer that is in a gaseous state, enveloping the Earth. If we analyse its mass, 99% of the atmospheric mass is concentrated in the first 30 km high section (equivalent to 0.5% of the Earth's radius). In the atmosphere, there are a series of gases, which together are known as air. In addition, the air contains solid and liquid particles (aerosols) in suspension, which can be found in variable composition.

The gases that make up the atmosphere are classified as follows:

- Permanent gases: their proportion remains practically constant. These are nitrogen (N_2), oxygen (O_2), noble gases (helium, neon, argon and krypton) and hydrogen (H_2).
- Gases in varying proportions: carbon dioxide (CO_2), water vapour (H_2O) and ozone (O_3).

The main particles, solid and liquid, that we find are helium crystals and water microdroplets (clouds, haze and fog). Table 1 below shows a classification of the main components of the air per unit of mass and volume, valid for altitudes below 100 kilometres.

Table 1. Atmospheric composition

Composition of the Atmosphere			
Gas	Symbol	% by weight	% by volume
Nitrogen	N_2	75.52	78.09
Oxygen	O_2	23.15	20.95
Argon	A	1.28	0.93
Carbon dioxide	CO_2	0.046	0.035
Neon	Ne	0.012	0.0018
Helium	He	0.0007	0.0005
Methane	CH_4	0.0008	0.00015
Krypton	Kr	0.003	0.0001
Ozone	O_3	0 – 0.01	Variable
Water vapour	H_2O	0 - 4	Variable

The air temperature varies with altitude. This is what is defined as the thermal gradient (dT/dz). It is the variation of air temperature (T) with height (z). The atmosphere is divided into layers, in vertical structure, according to the temperature gradient. There are three main zones:

a) Lower atmosphere or Homosphere: practically constant air composition. Up to a height of between 80 and 100 kilometres. This, in turn, is divided into three sub-layers:
   i. Troposphere: it is the lowest part of the atmosphere, where meteorological phenomena take place and develop. It contains approximately 80% of the total atmospheric mass. The temperature profile decreases as altitude is gained according to a thermal gradient of -6.8ºC/km. However, this profile is highly variable depending on the place and time. The upper part of this sub-layer is known as the tropopause, separating the troposphere from the stratosphere. Its height is variable, ranging from 6-8 kilometers at the poles to 16-18 kilometers at the equator.
   ii. Stratosphere: unlike the troposphere, the temperature profile increases with altitude (thermal inversion) which results in great stability since vertical movements are slowed down by this temperature inversion. The main radioactive energy exchanges take place in this sub-layer, as well as the phenomenon of absorption of ultraviolet radiation due to the presence of ozone (between 8 and 30 kilometres above sea level). The upper part of this sub-layer is called the stratopause and is about 50 kilometres high.
   iii. Mesosphere: the temperature profile decreases again with altitude in this sub-layer, where it reaches its lowest values (around -90ºC) in the upper part known as the mesopause.
b) Heterosphere: this layer begins at a height of between 90 and 100 kilometers. It is an atmosphere of low density and highly variable composition produced by chemical reactions and the phenomenon of gas diffusion due to the effect of gravity. The heterosphere is composed of:
   i. Thermosphere: again, the temperature profile is again increased by the height, reaching values ranging from 500 to 2000K in its upper part, depending on solar activity. The upper limit of the thermosphere is known as thermopause, whose height ranges

between 200 and 500 kilometers (very variable), depending on the same solar activity.

ii. Metasphere: sub-layer found at altitudes above 500 kilometres. It is also known as the magnetosphere because the movement of particles in this part of the atmosphere is strongly conditioned by the earth's magnetic field.

c) Exosphere: this is the highest or outermost layer of the atmosphere. The gas that is housed is of very low density. Ionized particles. It is called the set formed by the heterosphere and the exosphere as high atmosphere.

Table 2 shows the atmospheric structure or layers classified according to the vertical thermal gradient.

Table 2. Thermal gradient by atmospheric structure

Feature	Height (km)	Height (km)	Gradient
Troposphere	0	10	dT/dz < 0 (adiabática)
Tropopause	10	20	dT/dz = 0 (isoterma)
Stratosphere	20	40	dT/dz > 0
Stratopause	40	50	dT/dz = 0
Messosphere	50	80	dT/dz < 0
Mesopause	80	90	dT/dz = 0
Thermosphere	90	100	dT/dz > 0

Other areas of special interest that make up the atmosphere are:

- Ionosphere: this is a region between 60 and 600 kilometres above sea level. In this high part of the atmosphere, the components of the air are highly ionized due to the action of shortwave UV rays, X-rays and cosmic radiation. This region plays a fundamental role in the transmission of radio waves.
- Ozonosphere: located in the stratosphere, between 15 and 40 kilometres above sea level, this section is characterised by a high concentration of ozone, whose characteristic property of absorbing UV radiation prevents most of it from reaching the earth's surface. It acts as a protective shield against UV radiation.

For reference, the conditions of a standard atmosphere are defined:

- Air (% volume): 78.04% N_2, 20.99% O_2, 0.94% Ar and 0.035% CO_2.
- Standard conditions (sea level, z = 0):
  - Pressure 1,013.25 mbar.
  - Temperature 15ºC.
  - Density 1.2257 kg/m^3.
  - Acceleration of gravity 9.80665 m/s^2.

According to the concepts described above, a process in which there is no heat exchange with the surrounding medium is known as an adiabatic process. In the atmosphere, the relationship between the temperature profile that corresponds to or results from an adiabatic rise of the air and the actual temperature profile defines what is called the vertical stability of the atmosphere.

In terms of wind energy, the troposphere is the layer of greatest interest. In this sub-layer, the temperature profile for adiabatic stratification is characterized by a theoretical thermal gradient for dry air of -9.8ºC/km. Actually, as already mentioned, the value in practice and due to the presence of water vapour is reduced to -6.8ºC/km.

The phenomena of heating and cooling of the ground by solar radiation together with the mixture of air masses of different origins, results in the variation of temperature with altitude. This variation causes and conditions the vertical movements of the air. An atmosphere is considered stable if the vertical movements are inhibited. Conversely, instability facilitates such movements. A neutral atmosphere is indifferent to these movements. The following cases are considered when an air mass rises:

- Neutral atmosphere: the vertical temperature profile is such that, as one ascends, a neutral atmosphere presents the same temperature as it would have if the elevation were carried out adiabatically. In this case, the adiabatic temperature profile coincides with the actual temperature profile of the neutral atmosphere.
- Stable atmosphere: the vertical temperature profile is such that as it rises, the stable atmosphere has a higher temperature than it would have if the rise were adiabatic. The real temperature profile has a steeper slope than the adiabatic profile of dry air so that, if an air particle were to rise, its adiabatic temperature would be lower than the real temperature of the surrounding air so that the particle would tend to descend, inhibiting vertical movements.

- Unstable atmosphere: the temperature profile is such that, as one ascends, the unstable atmosphere has a lower temperature than it would have if the elevation were adiabatic. If an air particle were to rise adiabatically, its adiabatic temperature would be higher than the real temperature of the air, so it would tend to continue rising, creating vertical movements that would favour mixing in the atmosphere. The real temperature profile has a lower slope than the adiabatic profile of dry air.

In general, during the day, due to the radioactive action of the sun, the air masses warm up and the atmosphere tends to become unstable with a consequent increase in the intensity of the surface wind. On the other hand, during the night, they tend to stabilize.

The movement of air takes place mainly in the troposphere and this movement is influenced by the following factors:

- Solar radiation, higher in the equatorial zone than in the polar zone.
- Rotation of the Earth, produced by the Coriolis Effect, diverting the direction of the winds to the right in the northern hemisphere and to the left in the southern hemisphere.
- Action of the pressure gradient on the air masses, in addition to different types of surfaces (continental, maritime) and the orography of the terrain.

In the equatorial zone, motivated by the greater solar incident action, an ascending convective movement of the air masses is originated in the intertropical convergence zone. The air rises and moves towards higher latitudes, being replaced by the arrival of colder air from the tropics (trade winds). This phenomenon gives rise to Hadley cells. A similar process occurs in the northern latitudes giving rise to polar cells or Ferrel cells.

At mid-latitudes, the winds are basically westerly, but with a tendency towards the polar areas. For a latitude close to 60º there is a separation between the warm tropical air and the cold polar air, forming a polar front, with many undulations, giving rise to frontal depressions. In polar areas, cold air tends to move to lower latitudes.

In each hemisphere, there are three main nuclei: tropical, temperate and polar.

- In equatorial latitudes, the depression resulting from the warming of the lower layers of the troposphere exerts a suction effect of air in both hemispheres, in the intertropical convergence zone, causing on both sides of this zone the trade winds of east component, of regular character and with direction towards the equator. They reach 30º north and south latitude. The temperate nuclei are separated from the tropical ones by the so-called high-pressure zones, around 30º latitude. These zones are characterized by scarce precipitations and of general calm. In them we find the largest deserts such as the Sahara, Arabia and Gobi (northern hemisphere), Kalahari and Australia (southern hemisphere). From these calm zones diverge both trade winds and westerly winds.
- In temperate zones, which range from the tropics to the polar regions, the prevailing winds are from the west, more variable than the trade winds and modified by the presence of the continents. Air masses of both tropical and polar origin enter these regions, giving rise to successive warm and cold fronts with the appearance of clouds and precipitation.
- In the polar zones, which cover latitudes from 60º latitude onwards, they are characterized by high pressure and low precipitation. From these areas a flow is generated, with an approximate easterly direction, towards the low-pressure zone located at latitudes close to 60º.

In summary, the atmospheric circulation establishes, on a global terrestrial scale, a set of dominant wind directions as shown in Table 3:

Table 3. Dominant wind directions

Latitude range	90 - 60º N	60 - 30 ºn	30 - 0º N	0 - 30 ºS	30 - 60 ºS	60 - 90 ºS
Dominant direction	NE	SO	NE	SE	NO	SE

Wind is defined as air that is in a state of motion. As previously mentioned, on a global scale, it is the difference in temperatures caused by solar radiation over the different areas of the Earth that generates the wind. The differences in pressure and density, originally also generated by these temperature differences, are the cause of the movement of the air masses. The following scales (Table 4) are considered in the analysis or study of wind:

- Large-scale wind circulation is caused by pressure gradients in macroscale systems (anticyclones and squalls) and can last for several days.
- Mesoscale circulation is caused by pressure gradients in medium-scale systems (fronts, storms, breezes, etc...). Its duration can reach several hours. Mesoscale and microscale studies are the most interesting for the study of wind energy exploitation.

Table 4. Wind scales

Scale	Dimensions (km)	Example
Planetary	5000	Hadley cells
Macroscale	1000	Mid-latitude cyclones
Mesoscale	100	Storms and Brezzes
Microscale	10	Turbulence

Therefore, the wind is the result of the action of forces of the following origin: gravity, great pressure, Coriolis Effect and friction force.

The pressure gradient is what creates the movement of air masses. The greater the gradient, the greater the wind velocity. That is, the closer the isobar lines are (lines that represent points of the same pressure), the greater the intensity (velocity) of the wind.

In ideal atmospheric conditions, that is to say, without friction, due to the effect of the Coriolis force, the wind moves parallel to the isobars in such a way that, in the northern hemisphere, it leaves the high-pressure zone (anticyclone) to its right and the low-pressure zone (depression) to its left. The situation will be reversed in the southern hemisphere, where the baric force is balanced by the Coriolis force.

In the vertical structure of the troposphere, there is a difference between the layers closest to the Earth's surface, where the friction force is important, and those farther away, where it is considered insignificant (free atmosphere). In the first zone affected by friction, called the boundary layer, the friction force causes the wind direction component towards low-pressure zone to be greater and the wind velocity to be clearly different from the geostrophic wind value. Although the atmospheric boundary layer (ABL) has a variable height depending on the location, it is assumed that its mean height is 1.000 meters.

At heights above the boundary layer, the effects of friction are not very important and can be neglected in a first approximation of analysis. At this height level, the real wind velocity is approximately that of a geostrophic wind.

In wind energy applications, given that the height of energy capture is in the order of 100 meters, we always work within the boundary layer, a height where the effect of friction and the orographic profile of the terrain is very important.

In summary, the forces that cause the velocity at which the wind moves are:

a. Force due to the horizontal pressure gradient. It is directed perpendicularly from higher to lower pressure isobars.
b. Force due to the Coriolis effect: rotation of the Earth from west to east, whereby the trajectory of a moving air mass is deflected to the right in the northern hemisphere and in the opposite direction in the southern hemisphere.
c. Centrifugal force: due to the curvature of the isobars, it acts in a direction perpendicular to the radius of curvature of the trajectory.
d. Frictional or frictional force: it intervenes in the lower layers of the atmosphere, close to the earth's surface. Factors such as the orography of the terrain or the type of surface of the terrain influence the effect of this force on the wind.

According to the types of forces considered, the following types of wind are classified:

i. Geostrophic wind: resulting from the joint action of the forces due to the pressure gradient and the Coriolis Effect. Wind parallel to the isobars, being higher the closer the isobars are to the one with the highest pressure. For the same value of pressure gradient, it decreases with increasing latitude. An observer located in the northern hemisphere with his back to the wind leaves the centres of low pressure (depressions or squalls) to his left and those of high pressure (anticyclones) to his right.
ii. Grandient wind: resulting from the forces generated by the pressure gradient, the Coriolis Effect and the centrifugal force. The force caused by the pressure gradient is directed from the higher-pressure isobars to the lower pressure isobars perpendicularly. The centrifugal force is directed towards the convex part of the trajectory. Therefore, in an anticyclone situation the centrifugal force is added to the force resulting from the pressure gradient, while if it produces a depression or squall it is subtracted. Therefore, for the same pressure gradient, in

the anticyclone zone the geostrophic wind velocity is lower than that of the gradient wind. This type of wind is very representative in situations where the friction force is negligible, occurring in the layers far from the earth's surface.

For these reasons, geostrophic and gradient winds describe with a very close result the actual wind conditions above 1.000 meters above ground level. At lower altitudes, up to about 100 meters, the wind is strongly influenced by the orographic profile of the terrain, terrain roughness and obstacles. In wind energy applications, these are the types of wind behaviour to be analysed, i.e., wind behaviours in the atmospheric boundary layer.

i. Cyclotrophic wind: occurs on a smaller scale (mesoscale), such as tornadoes or the center of a tropical cyclone. The strength of the Coriolis effect plays no role. When a phenomenon such as a tornado is formed, both the wind velocity and centrifugal force increase faster than the Coriolis force due to the large force generation by pressure gradient, so the centrifugal force is balanced by the pressure gradient in this case. This type of wind can occur in both hemispheres as a result of the independence of the Coriolis force in a clockwise or counterclockwise direction.

The geostrophic wind direction is parallel to the isobars. Due to friction, near the ground surface, the real wind slows down and changes its direction so that it cuts the isobars in the direction of the low-pressure zones. This effect of friction or friction of the ground is dependent not only on the relief but also on the type of weather.

For moderate winds and cloudy skies, the air velocity increases from a zero value at ground level to its geostrophic value at a height of about 500 to 1.000 meters, rotating clockwise with height. In summer, during the midday period when radiation is at its maximum, the effect of friction or friction can be felt up to a height of 1.000 to 2.000 metres, intensifying the phenomenon of convection that causes the air to rise from the ground. At night, if it is particularly clear, the wind in the lowest part of the atmosphere is almost nil and is only accelerated by the gravitational effect that flows towards the valleys.

As mentioned above, on mean, it is considered that the effects of friction or friction present on the earth's surface manifest themselves up to a mean height of about 1.000 metres. This is what is known as the boundary layer.

Outside the boundary layer, geostrophic wind and gradient conditions are approaching.

Within the boundary layer, we differentiate the layer closest to the ground or surface layer, up to about 100 meters high, the Elkman layer, where changes in wind direction and velocity occur more progressively and smoothly [1-6].

Methods

Wind Data Treatment. Weibull´s Distribution Law

An analysis for the study of the viability of wind as an energy source requires knowledge of its behaviour. As already described, the physically probable areas for the location of wind power generation facilities are close to the most optimistic boundary layer values on the earth's surface. For their variable nature, statistical analysis techniques have to be applied. The following processes are involved:

- Data collection of wind velocity, direction. It is very important to establish a measurement procedure.
- Application of measurement techniques with the appropriate instrumentation under the previously detailed measurement procedure.
- Application of methodologies such as statistical techniques, graphical presentation of results, calculation of probability distributions, etc.

Attention should be paid to the following aspects:

- Statistical analysis of wind velocity and direction.
- Temporal distribution of velocity and direction.
- Weather conditions.
- Vertical profile of ground velocity.
- Turbulence.

The collected wind data are treated statistically in order to determine the frequency distribution of the velocities and the directional distribution of the wind direction.

The applied methodology is based on the statistical analysis of the wind data collected for the subsequent application of the Weibull distribution method. The Weibull distribution law of wind velocity allows to calculate its energetic potential and the energy produced by a wind turbine over a period of time, so the results can vary depending on the time series and data used. This law represents the behaviour of the wind velocity at a specific location and height. In a complementary way, energy production parameters of the wind turbine and of the complete set will be calculated based on the initial data. This method represents a preliminary calculation tool whose objective is to carry out an evaluation of the site's potentiality, analysis for the optimal choice of the wind turbine as well as to build a base on which to carry out an economic and feasibility study of execution.

The Weibull distribution is a versatile distribution that can be used to model a wide range of applications in engineering, medical research, quality control, finance, weather forecasting and other examples. Such a distribution is described according to the parameters or factors of shape, scale and threshold and termed as the three-parameter Weibull distribution. In the case where the threshold parameter or factor is zero it is known as the two-parameter Weibull distribution. The two-parameter Weibull distribution is defined only for positive variables, as opposed to the three-parameter Weibull distribution, which can work with zeros and negative data.

- EParameter effect or shape factor: describes the way the data is distributed. A shape factor of 3 closely approximates a normal curve. A low shape factor defines a curve with an asymmetrical shape towards the right side. A high shape factor defines a curve that is asymmetrically shaped to the left.
- Parameter effect or scaling factor: defines the position of the Weibull curve with respect to the threshold value, which is similar to the way the mean defines the position of a normal curve.
- Effect of the threshold value parameter or factor: Describes a shift of the distribution away from 0. A negative threshold value shifts the distribution to the left while a positive threshold value shifts the function to the right. All data must be greater than the threshold value.

The complete methodology, developed in the following sections, aims to calculate the energy potential of an area of possible location for a wind farm from a series of data that include: wind metrics performed in that location in addition to technical parameters of the wind turbine, also object of evaluation to optimize the model to choose.

Usually, wind resource studies and energy potential calculations are sufficiently representative when data series of 5 to 10 years, or even more, are grouped together. However, the described methodology analyses measurement variables in the one-year time series for a particular measurement point on the potential site. As will be described the method allows the addition or modification of further time series, as well as to evaluate different possibilities with respect to the wind turbine. The objective is to proceed with a basis that will run a preliminary automatic analysis of the energy potential of a location from an initial measurement point and a time series of collected data. The development of improvement of this procedure, i.e., the robustness of this procedure, involves adding more data to the time series or modifying the time series, in which case the modifications are of a minimal nature.

There are a number of parameters or variables, namely statistical concepts, which are used in the analysis of wind studies. Although it is obvious that the most known and used variable is the velocity, there are also other concepts that describe this phenomenon, such as wind direction. We divide these variables into two types:

Centralization Parameters

On a set of N measured values of wind velocity ($v_1 \ldots v_i$), the following parameters are defined:

- Equation 1 shows Relative frequency f_i of the quantity v_i:
  $$f_i = \frac{n_i}{N} \tag{1}$$
- Euation 2 shows Arithmetic mean or mean (v):
  $$\langle v \rangle = \sum_{i=1}^{k} \frac{n_i v_i}{N} = \sum_{i=1}^{k} f_i v_i \tag{2}$$
- Median ($\tilde{v}$): the middle value of the two values that divide a set of data ordered in magnitude. Those values that divide the data set into four equal parts are called quartiles ($v_{0.25}$), if they do so into ten equal parts, they are called deciles ($v_{0.10}$) and if they do so into one hundred equal parts they are called percentiles ($v_{0.01}$). The median is the second quartile, the fifth decile and the fiftieth percentile.
- Mode (v_m): value of the data set that occurs with the maximum frequency.

Dispersion Parameters

In a data set, measures of dispersion describe the distribution around a mean value. The most commonly used are:

- Range: difference between the largest and smallest of the values.
- Range between percentiles: usually the ranges 10-90, 5-95 and 1-99 are used, which represent 80%, 90% and 98% of the dataset. These ranges are calculated as the differences between the percentile $v_{0,90}$ and percentile $v_{0,10}$, etc...
- Equation 3 shows Standard deviation or typical deviation (σ): in a set of N wind velocity data (v 1, ..., v i) described by the relative frequencies (f 1, ..., f i), it is calculated as:

$$\sigma = \sqrt{\sum_{i=1}^{k} f_i (v_i - \langle v \rangle)^2} \tag{3}$$

The expression may be described as equation 4 shows:

$$\sigma = \sqrt{\sum_{i=1}^{k} f_i v_i^2 - \left(\sum_{i=1}^{k} f_i v_i\right)^2} = \sqrt{\sum_{i=1}^{k} f_i v_i^2 - \langle v \rangle^2} \tag{4}$$

- Variance (σ^2): defined as the square of the standard deviation.

These parameters give us the description and parameterization of the wind velocity data set (in this case) that is taken as a basis for the development of the procedure. On this set of data, additionally, a statistical treatment is carried out for the construction of a table and figures that describe and graphically show the data collection carried out. The processing of this set of N velocity values (v_1, ..., v_i) consists of the following steps:

- The process of grouping the data set into classes, determining the frequency of each class. That is, the amount of data belonging to each class.
- The number of classes to be determined depends on the number of data collected, the range of the complete set of data taken, or the precision in the calculation for exigency or decision reasons.
- Construction of a table including the determined classes and the calculated frequencies. Then, the relative frequencies, cumulative

frequencies, mean velocity, median, mode and standard deviation are calculated.

- Plot the relative frequency histogram, the cumulative frequency curve, either the "less than" or its complementary "greater than".
- Represent the velocity duration curve, which describes the time during which the velocity remains above a specific value.

Table 5 shows the grouping and construction made on the basis of "type" wind velocity measurements taken. Thirty-one wind velocity classes have been taken for hourly wind velocity measurements during the period of one year. That means we start from 8760 data describing wind velocity. These data would be those taken at the site under study in the initial period. Dispersion and centralization parameters have been calculated based on the described procedure.

Additionally:

- Figure 1 shows the histogram of relative frequencies.

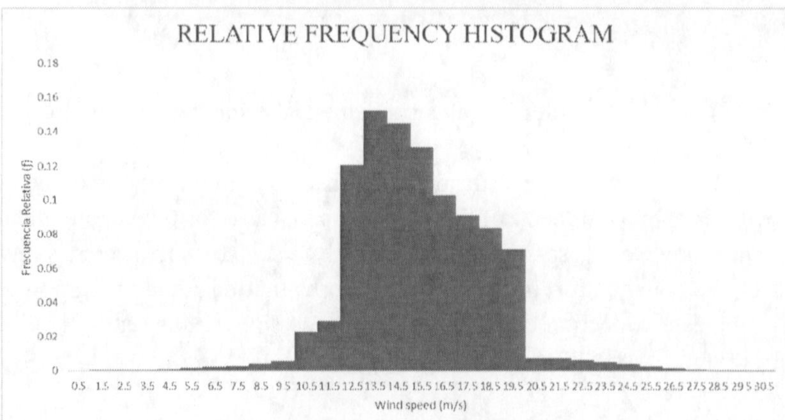

Figure 1. Relative Frequency Histogram.

- Figure 2 shows the "greater than" cumulative frequency curve.

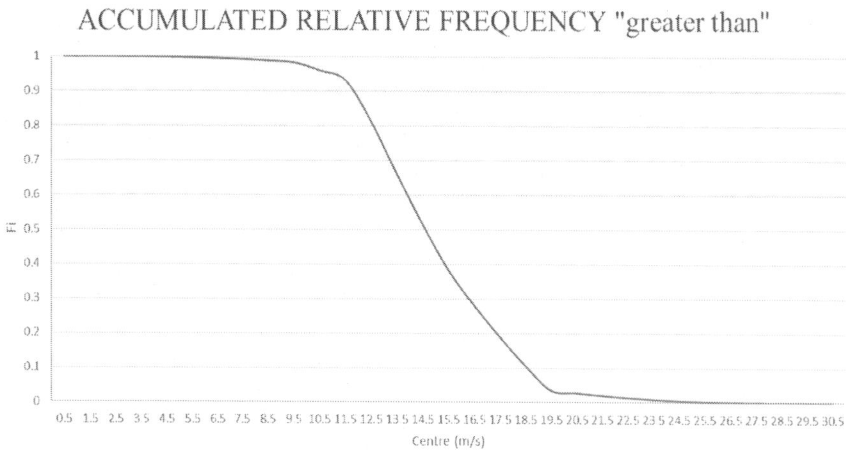

Figure 2. Accumulated Relative Frequency "greater than".

- Figure 3 shows the "less than" cumulative frequency curve.

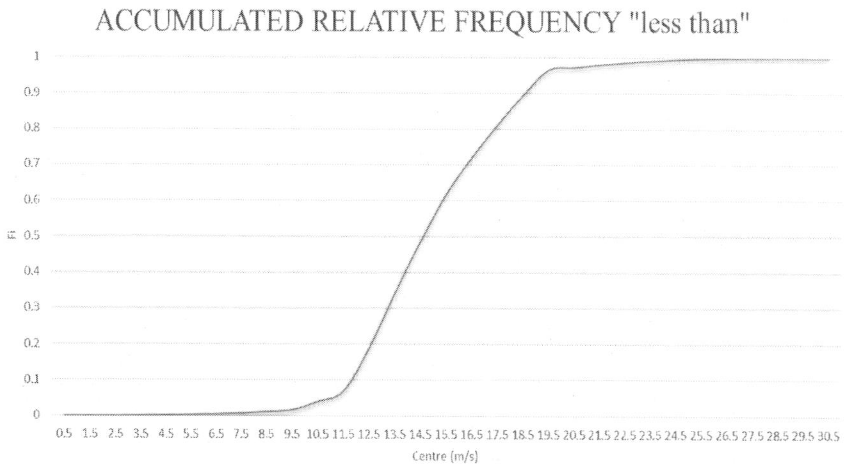

Figure 3. Accumulated Relative Frequency "less than".

- Figure 4 shows the velocity duration curve for the data set.

As already mentioned, for wind energy studies, the data set of a single year in most cases is not sufficiently representative. Before even a measurement campaign is carried out where some of the wind turbines are planned to be placed and relate the new measurements taken with the previous statistical treatment and possible modelling studies and / or wind behaviour. What is certain, however, is that regardless of the period of data collection, the histograms resulting from the process correspond to a Weibull distribution.

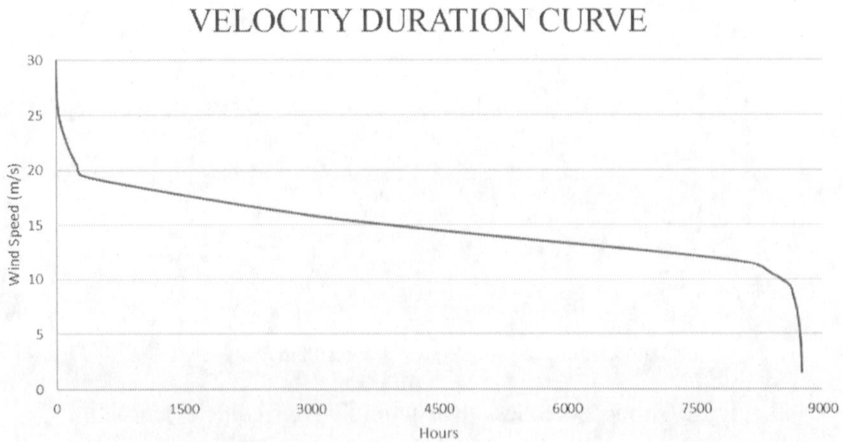

Figure 4. Velocity Duration Curve.

Weibull´s Distribution Law. Definitions

The wind velocity distribution follows, for most cases of interest in wind energy, the Weibull probability density function p(v). Figure 5 shows a real distribution of measured velocities and the fitted Weibull law. This function is calculated as equation 5 shows:

$$p(v) = \frac{k}{c}\left(\frac{v}{c}\right)^{k-1} e^{-\left(\frac{v}{c}\right)^{k}} \tag{5}$$

Where:

- v: wind velocity or velocity (m/s).
- p(v): Weibull probability density function.
- c: scale factor (m/s). Variable of value close to the annual mean velocity. Later it will also be described as parameter A.
- k: shape factor that gives bias to the function.

Properties of Weibull's Law

The n-th moment of the Weibull distribution is given as equation 6 shows:

$$\int_0^\infty v^n\, p(v)\, dv = c^n \Gamma\left(1 + \frac{n}{k}\right) \quad (6)$$

The probability density has a maximum or mode for the velocity v_m is given by equation 7:

$$\frac{v_m}{c} = \left(\frac{k-1}{k}\right)^{1/k} \quad (7)$$

The probability that the wind velocity is less than or equal to a certain value V, is given by the cumulative distribution F, defined by equation 8:

$$F(v \leq V) = \int_0^v p(v)dv = 1 - e^{-\left(\frac{v}{c}\right)^k} \quad (8)$$

The complementary cumulative distribution F' allows calculating the cumulative frequency of velocities that are greater than or equal to a certain V value as equation 9 shows:

$$F'(v \geq V) = 1 - F(v \leq V) = e^{-\left(\frac{v}{c}\right)^k} \quad (9)$$

The probability that the velocity v is between two values v_1 and v_2 is shown by equations 10 and 11:

$$P(v_1 \leq v \leq v_2) = \int_{v_1}^{v_2} p(v)dv = F(v_2) - F(v_1) = F'('v_1) - F(v_2) \quad (10)$$

$$P(v_1 \leq v \leq v_2) = e^{-\left(\frac{v_1}{c}\right)^k} - e^{-\left(\frac{v_2}{c}\right)^k} \quad (11)$$

Statistical Parameters of the Velocity Distribution

For a Weibull distribution, the annual mean velocity, median and variance are calculated according to the following expressions:

$$\langle v \rangle = \int_0^\infty v\, p(v)dv = c\,\Gamma\left(1 + \frac{1}{k}\right) \quad (12)$$

$$\tilde{v} = c\,(ln2)^{1/k} \tag{13}$$

$$\frac{\sigma^2}{c^2} = \Gamma\left(1 + \frac{2}{k}\right) - \left[\Gamma\left(1 + \frac{1}{k}\right)\right]^2 \tag{14}$$

Where r(x) is gamma function, defined by equation 15:

$$\Gamma(x) = \int_0^\infty e^{-z}\, z^{x-1} dz \tag{15}$$

Energy Parameters

One is the annual mean of the cube of the velocities, which should not be used incorrectly by applying the cube of the annual mean velocity. For a Weibull distribution, it is calculated as equation 16 shows:

$$\langle v^3 \rangle = \int_0^\infty v^3 p(v) dv = c^3 \Gamma\left(1 + \frac{3}{k}\right) \tag{16}$$

The effective velocity v* is defined as the velocity whose cubed annual mean of the cubes of the velocities is obtained. It is calculated as following equation:

$$v^* = \langle v^3 \rangle^{1/3} \tag{17}$$

The following is a list of parameters used in energy calculations:

- Annual mean velocity/scale factor c:
  $$\frac{\langle v \rangle}{c} = \Gamma\left(1 + \frac{1}{k}\right) \tag{18}$$
- Standard desviation/anual mean velocity:
  $$\frac{\sigma}{\langle v \rangle} = \left[\frac{\Gamma\left(1+\frac{2}{k}\right)}{\Gamma^2\left(1+\frac{1}{k}\right)} - 1\right]^{1/2} \tag{19}$$
- Energy pattern factor o Wind power factor (F_e):
  $$F_e = \frac{\langle v^3 \rangle}{\langle v \rangle^3} = \frac{\Gamma\left(1+\frac{3}{k}\right)}{\Gamma^3\left(1+\frac{1}{k}\right)} \tag{20}$$

This factor varies in most cases between 1.5 and 3 and is generally set at 2.

- Effective velocity/Annual mean velocity:

$$\frac{v^*}{\langle v \rangle} = F_e^{1/3} \tag{21}$$

- Median/Annual mean velocity:

$$\frac{\tilde{v}}{\langle v \rangle} = \frac{(ln2)^{1/k}}{\Gamma\left(1+\frac{1}{k}\right)} \tag{22}$$

Weibull Parameters. Distribution Weibull Curve: Calculation

For the calculation of the Weibull parameters (k, c) two types of methodology are presented depending on the known starting data [7].

Starting from the Measured Minute, Ten-Minute, Half-Hourly or Hourly Velocities for One Year: Least Squares Adjustment Method

In this case, the starting point is the hourly wind velocity (hourly measurements). The parameters k and c can be determined by least squares fit through the cumulative distribution F(v) as equation 23 shows:

$$F(v) = 1 - e^{-\left(\frac{v}{c}\right)^k} \tag{23}$$

Applying neperian logarithms to the previous expression, we obtain as following equation:

$$\ln(1 - F(v)) = -\left(\frac{v}{c}\right)^k \rightarrow ln[-\ln(1 - F(v))] = k\ lnv - k\ lnc \tag{24}$$

Next, we proceed to the least squares fit to the line y = Ax+B, where:

$$y = ln[-\ln(1 - F)]\ ;\ x = lnv\ ;\ B = -k\ lnc \tag{25}$$

Calculated the line by least squares, the slope gives the value of k and the value of c is calculated as equation 26 shows:

$$c = e^{-\left(B/k\right)} \tag{26}$$

With this method of adjustment, the Weibull parameters k and c and, therefore, the probability density function p(v) are determined. The degree of accuracy will depend, as mentioned above, on the amount of data available.

Based on the Annual Mean Velocity and the Standard Deviation

From the standard deviation/mean annual velocity ratio, the value of k is obtained, to then calculate the mean annual velocity parameter/scaling factor and finally to obtain the parameter c from this same expression. This process requires the resolution of a transcendent equation with the gamma function. Therefore, it is preferable to use one of the following methods:

- Interpolate in the table included in the Annex, obtaining the value k of the relationship Standard deviation/annual median velocity. After calculating k, the parameter c is obtained following the process indicated in the previous paragraph.
- Using an approximate expression for the calculation of k, parameter c is calculated by applying the procedure in the previous paragraph. The expression is calculated as equation 27 shows:

$$k = 0.9791\left(\frac{\langle v\rangle}{\sigma}\right)^{1.1058} \cong 0.98\left(\frac{\langle v\rangle}{\sigma}\right)^{1.11} \quad (27)$$

For our particular applied case, we are going to apply option a). We already have the starting data, as Table 1 shows, on which the least squares adjustment will be made. The expressions used to calculate the coefficients of the linear regression line is described by expression 28:

$$A = \frac{\sum f_i x_i y_i - (\sum f_i x_i)(\sum f_i y_i)}{\sum f_i x_i^2 - (\sum f_i x_i)^2} \quad ; \quad B = \sum f_i x_i - A \sum f_i x_i \quad (28)$$

Once the adjustment is applied, the Weibull parameters are obtained for our specific case of application, a measurement point located 60 meters above sea level in a specific, defined location. The parameter k obtained is 5.79 and the parameter c obtained is 16.03. These are the parameters that are define in Table 6.

Complementarily, the calculation and representation of the distribution curve of the probability function p(v) and cumulative frequency F(v) that represent and describe the initial data set is carried out. For this purpose, the calculation has been carried out using a “manual” methodology by calculating the different parts of the Weibull function and operating them and, on the other hand, the Weibull function has been applied in MS Office Excel for checking and comparing results.

Finally, it is necessary to take into account which model of wind turbine will be the reason for the analysis of the study for the production of electrical energy. In any wind turbine that is going to be the subject of analysis, it is necessary to take into account the height at which the hub of the same is located, since this is the point on which it has to be taken as a reference to obtain the correct k and c parameters. In this case, the “type” wind turbine will be the Gamesa SG 5.0-132 model, with a maximum power of 5 MW, a hub height of 84 metres and a swept area of 13.685 m^2. It is necessary to obtain the new parameters of the Weibull function for this new reference height [8].

Knowing the Weibull parameters for a known height (60 meters in this case), the Justus and Milhail expressions have developed the approximate empirical relationship with respect to the new values for the new calculated height. Note that their use is not recommended for heights greater than about 100 meters. These are decribed by following equations:

$$k' = k\left(\frac{1-0.088\,ln\left(\frac{z}{10}\right)}{1-0.088\,ln\left(\frac{z'}{10}\right)}\right) \quad (29)$$

$$c' = c\left(\frac{z'}{z}\right)^{\beta} \quad (30)$$

Where the coefficient β is calculated as equation 30 shows:

$$\beta = \frac{0.37-0.088\,ln c}{1-0.088\,ln\left(\frac{z}{10}\right)} \quad (31)$$

For clarification:

- k – k′: shape parameters for the heights z y z′respectively.
- c – c′: scale parameters for the heights z y z′ respectively (m/s).

Again, these are approximate expressions, recommended for terrains with low roughness and in the case of not having measurements of one hundred for the desired height, as in our case.

Applying the training described above, new parameters of scale and shape will be obtained and, therefore, a new Weibull distribution for this new reference point. The new parameters obtained, at 84 meters height, are 6.00 for the k parameter and 16.15 for the c parameter.

Additionally, through Hellman's law, the variation of the mean annual velocity with height or the new mean annual velocity (v') is calculated, a necessary parameter for the development of the following section, according to the following expression:

$$\frac{\langle v' \rangle}{\langle v \rangle} = \left(\frac{z'}{z}\right)^{\alpha} \tag{32}$$

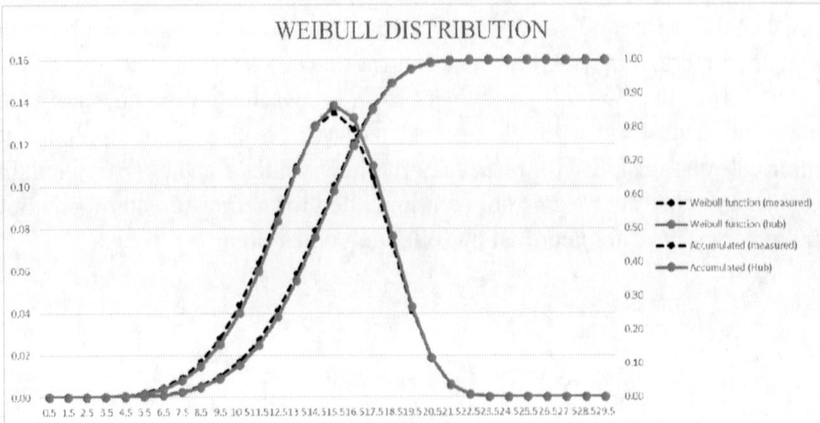

Figure 5. Weibull distribution (measured and hub height).

In our case, this parameter will be calculated using the "gamma" function in MS Office Excel.

In conclusion, Figure 5 represent the Weibull probability function p(v) for each height (measured and bushing height) in addition to the cumulative frequency F(v).

Wind Power Potential: Calculation Method

Wind potential describing an area is based on the principle of a mass flow rate that travels at a specific air surface perpendicular to the direction of flow, i.e., by its kinetic energy, which is calculated as:

$$E_c = \frac{1}{2} m v^2 \tag{33}$$

Mass flow rate of air (m) of density (ρ) flowing with velocity (v) across such an area surface (A) perpendicular to the direction of flow is calculated as:

$$m = \rho\, A\, v \tag{34}$$

Available wind power (P_d) of the airflow through this section is:

$$P_d = \frac{1}{2}\, m\, v^2 = \frac{1}{2}\rho A v^3 \tag{35}$$

Available wind power is proportional to the air density, to the area perpendicularly exposed to the wind flow and to the cube of its velocity. If this were maximum it would mean that we are extracting all the kinetic energy of the wind. Real cases do not exceed 45-50%, due to other technical, mechanical and/or electrical considerations. Therefore, the available power density is:

$$\frac{P_d}{A} = \frac{1}{2}\rho v^3 \tag{36}$$

Due to the nature of velocity variability, definition of the available wind potential at a site has to be obtained by determining power values and averaging them over a period. In this way, the wind potential at a reference point is defined as the mean wind power per unit area and period, assuming constant air density. It is calculated as:

$$\frac{\langle P_d \rangle}{A} = \frac{1}{2}\rho \sum_0^\infty v^3\, p(v)dv = \frac{1}{2}\rho\langle v^3\rangle = \frac{1}{2}\rho F_e \langle v\rangle^3 \tag{37}$$

For the specific Weibull distribution, applicable to wind energy, the probability density of the wind power density Φ(v) is calculated as:

$$\phi(v) = \frac{P_d}{A} p(v) = \frac{1}{2}\rho v^3 \frac{k}{c}\left(\frac{v}{c}\right)^{k-1} e^{-\left(\frac{v}{c}\right)^k} \tag{38}$$

This probability density function of the wind power density Φ(v) presents a maximum (mode) for a velocity v_{mp} different from the velocity v_m corresponding to the maximum of the probability density p(v). The velocity v_{mp} is obtained by calculating the maximum of the above expression. Deriving, equaling to zero and solving the equation we obtain:

$$\frac{v_{mp}}{c} = \left(1 + \frac{2}{k}\right)^{1/k} \tag{39}$$

The velocity v_m for which the maximum of the velocity distribution is given by the formula of the mode, therefore:

$$\frac{v_{mp}}{v_m} = \left(\frac{k+2}{k-1}\right)^{1/k} \tag{40}$$

Relation between the velocity v_{mp} and the annual mean is obtained as:

$$\frac{v_{mp}}{\langle v \rangle} = \frac{\left(1+\frac{2}{k}\right)^{1/k}}{\Gamma\left(1+\frac{1}{k}\right)} \tag{41}$$

For a Weibull distribution, the mean annual wind power density is:

$$\frac{\langle P_d \rangle}{A} = \frac{1}{2}\rho \sum_0^\infty v^3 p(v) dv = \frac{1}{2}\rho c^3 \Gamma\left(1 + \frac{3}{k}\right) \tag{42}$$

In general, the wind potential or available wind power is given by the following expression:

$$\frac{\langle P_d \rangle}{A} = \frac{1}{2}\rho \langle v^3 \rangle = \frac{1}{2}\rho F_e \langle v \rangle^3 = \frac{1}{2}\rho v^{*3} \tag{43}$$

On this expression, to be considered:

- [v*] represents the effective velocity, that is, the velocity that the wind should have constantly during the whole period of time to present a mean density of available power equal to the one it has in reality.
- [v^3] represents the mean of the cubes of the velocities in the corresponding period. Not to be confused with the cube of the mean velocity [v^3].

Available wind energy density (E_d) per unit area (A) is defined as the total energy of the wind at its direction during a given period of time (T). Its value is calculated as equation 44 shows:

$$\frac{\langle E_d \rangle}{A} = \frac{\langle P_d \rangle}{A} T \qquad (44)$$

It may also be calculated through the following expression:

$$\frac{\langle E_d \rangle}{A} = T \sum_0^{\infty} \left(\frac{P_d}{A}\right) p(v)dv = \frac{1}{2} T\rho \langle v^3 \rangle = \frac{1}{2} T\rho F_e \langle v \rangle^3 \qquad (45)$$

If statistical data in the form of frequencies (f_i) and velocities (v_i) are available, calculation is shown by equation 46:

$$\frac{\langle E_d \rangle}{A} = T \sum_i f_i \left(\frac{P_d}{A}\right)_i = \frac{1}{2} T\rho \sum_i f_i v_i^3 \qquad (46)$$

Continuing with the application of this specific case of taking measurements for a period of one year, we will proceed to calculate the power and wind energy density at the measured point and the point where the wind turbine hub would actually be placed, in both cases taking into account the use of the same wind turbine. To finalize the calculation procedure, using the wind turbine power curve and the Weibull distribution obtained, the power density (gross and net), energy density (gross and net) as well as the total production per wind turbine and for the installation are calculated, in this case taking into account that the calculation has been designed for 30 wind turbines. Finally, additional wind turbine operating parameters will be obtained. These are the mean global yield, the load factor, equivalent hours and the operation factor.

- Specific production or specific output (E/A): is the energy produced during a period of time T per unit area swept by the rotor of the wind turbine. This parameter represents the energy density that is harnessed by the wind turbine (technical limitations). In contrast to this density, this represents the electrical energy extracted from the available wind energy. Both parameters will be shown later.
- Overall annual mean electrical efficiency (η_{ea}): is the ratio between the electrical energy (E_e) produced in a period of time T and the available wind energy (E_d) in the same period. It may also be calculated as the ratio between the annual mean electrical power produced by the wind turbine (P_e) and the annual mean wind power available. It is calculated as:

$$\eta_{ea} = \frac{E_e}{E_d} = \frac{\langle P_e \rangle}{\langle P_d \rangle} \tag{47}$$

Mean available wind power (P_d) is calculated from the wind potential (P_d/A) by measuring, as already mentioned, the hub height and the swept area (A) by the rotor diameter (D) of the wind turbine:

$$\langle P_d \rangle = \frac{\langle P_d \rangle}{A} x\, A = \frac{1}{2}\rho A F_e \langle v \rangle^3 = \frac{1}{2}\rho A v^{*3} \tag{48}$$

- Load or capacity factor (FC): ratio between the electrical energy (E_e) produced in a period of time T and that would have been produced if, in the same period, it operated continuously at nominal power (P_N). It is calculated as:

$$FC = \frac{E_e}{P_N\, x\, T} = \frac{\langle P_e \rangle}{P_N} \tag{49}$$

An assessment may be made based on the result of this load factor. Based on empirical data, we can classify:

- Extraordinary: > 0.5
- Excellent: 0.4 - 0.5
- Very good: 0.3 – 0.4
- Good: 0.25 - 0.3
- Acceptable: 0.2 – 0.25
- Insufficient: < 0.22

Load factor increases with the mean wind velocity and thus with the annual energy produced. However, this is not the case for the yield. It is explained by the fact that the yield is a ratio between the energy produced and the energy that can be extracted from the wind. The first variable grows quasilinearly with the wind velocity and even stabilizes when the machine reaches its nominal power, while the second increases according to the cubic law and, therefore, more rapidly.

- Equivalent hours at full load (HE): number of hours that the wind turbine would have to operate at full load (nominal potential) to produce the same amount of energy annually. It is calculated as:

$$HE = \frac{E}{P_N} = \frac{\langle P \rangle T}{P_N} \tag{50}$$

- Operating factor (FO): in a period (T) is the ratio between the number of hours it has been producing electricity and the total number of hours in that period (T). It is calculated as:

$$FO = \frac{Running\, up\, annual\, hours}{8760\ hours/year} \tag{51}$$

Other parameters of recomendable knowledge are:

- Reliability: probability that the wind turbine will operate continuously and without failure, under specific conditions, over a period.
- Hours of unscheduled unavailability: annual hours of unscheduled downtime due to breakdowns or incidents (corrective maintenance, stop-start manoeuvres, etc…).
- Hours of scheduled unavailability: corresponds to the annual hours of scheduled shutdowns (preventive maintenance).
- Availability: probability that the wind turbine is in an operational state, under specific conditions. Currently this value, in modern wind turbines, is above 95%.

In order to calculate the wind power levels, energy density, Weibull distribution and annual production of the installation in the described assumption, the described calculations have been carried out. Prior to the presentation of results, some additional considerations have to be taken into account:

- A reference value of air density 1.225 kg/m^3 and a roughness length of the terrain 0.001 is taken, i.e., considering open and smooth terrain according to the list of roughness classes and lengths. The European Wind Map classifies the terrain into roughness classes according to the roughness length z0. This relation is expressed as:

$$If\ z_0 \leq 0.03\ m \rightarrow Class = \frac{ln(5000\ z_0)}{ln 150} \tag{52}$$

$$If\ z_0 > 0.03\ m \rightarrow Class = \frac{ln\left(\frac{1000}{9} z_0\right)}{ln\left(\frac{10}{3}\right)} \tag{53}$$

The list of roughness clases and lengths is classified below:

- Roughness class 0: up to 0.02 meters. Energy index 100%. Water surface. Open ground, smooth surface.
- Roughness class 0.5: up to 0.0024 meters. Energy index 73%. Concrete runways (airports), grass.
- Roughness class 1: up to 0.03 meters. Energy index 52%. Open fields without fences or hedges. Very scattered buildings. Gently rounded hills.
- Roughness class 1.5: up to 0.055 meters. Energy index 45%. Field with some houses and trees up to 8 meters located at least 1250 meters.
- Roughness class 2: up to 0.1 meters. Energy index 39%. Fields with some houses and trees up to 8 meters located at least 500 meters away.
- Roughness class 2.5: up to 0.2 meters. Energy index 31%. Field with many houses and trees up to 8 meters located at least 250 meters away.
- Roughness class 3: up to 0.4 meters. Energy index 24%. Villages, forests and rough and uneven terrain.
- Roughness class 3.5: up to 0.8 meters. Energy index 18%. Cities with tall buildings.
- Roughness class 4: up to 1.6 meters. Energy index 13%. Large cities with very tall buildings.
- Wind profile (coefficient α), a parameter that depends on the roughness of the terrain, has been calculated. There are different classification tables or lists for different terrains or depending on the length of the roughness of the terrain. For values of $z_0 \leq 0.1$ metres, as in this case, an approximate equation that can be used is:
  $$\alpha = 0.24 + 0.04\, lnz_0 + 0.003(lnz_0)^2 \quad (54)$$

In our case the result of the wind profile is 0,11. This is in line with what is expected, since we have assumed a flat terrain with reduced roughness.

- Electrical losses of 3%, losses due to wake effect of 2%, losses due to programmed unavailability of 3% and others due to incidents or programmed unavailability of 1% have been considered. In total, an additional loss coefficient of 9% will be applied.

Results

Parameters describing the energy potential of the measuring points and, most importantly, the computer bushing height, have been calculated.

Table 5. Parameters at measured height

Measured height	
k	*5.79*
c	*16.03*
Annual mean velocity (m/s)	14.84
Median	15.05
Variance	8.83
Standard desviation	2.97
Annual mean of velocity cubed	3,652.47
Effective velocity	15.40
Available Wind Power per area (W/m²)	2,237.1
Available Wind Energy density (MWh/m²)	19.6
WTG Power [MW]	30.6
WTG Energy density [MWh]	268,184.9

Table 6. Parameters at hub height

Hub height	
k	*6.00*
c	*16.15*
Annual mean velocity (m/s)	14.98
Median	15.19
Variance	8.43
Standard desviation	2.90
Annual mean of velocity cubed	3,730.17
Effective velocity	15.51
Available Wind Power per area (W/m²)	2,284.7
Available Wind Energy density (MWh/m²)	20.0
WTG Power [MW]	31.3
WTG Energy density [MWh]	273,889.9

- Table 5 and Table 6 show the following calculated parameters: mean annual velocity, median, variance, standard deviation, mean annual velocity of the cube of the velocities, effective velocity, available wind power of the site, available wind power at the site, available wind power based on the swept area of the wind turbine and available wind power based on the swept area of the wind turbine. Every parameter has been calculated at the point of measurement height and at the actual hub height of the wind turbine. The equations described above have been used.

Table 7 shows the parameters corresponding to the measuring heights and hub heights, technical parameters of the wind turbine, number of wind turbines for the case study, losses applied to the installation, as well as the parameters described above with respect to the terrain.

Table 7. WTG - Wind farm properties

WTG – Wind farm propertiesr	**Measured**	**Hub/Plant**	**Increase**
Height	60	84	40.00%
Mean velocity	14.84	14.98	0.93%
k-factor	5.79	6.00	3.64%
c-factor	16.03	16.15	0.73%
Power (W/m^2)	2,237.1	2,284.7	2.13%
Wtg Model	WTG		
Manufacturer	Gamesa		
Hub height (m)	120		
Spindle diameter (m)	132		
Number WTG	30		
Max Power (MW)	5		
Max Power Plant (MW)	150		
Swept area (m^2)	13,685		
Density	1.225		
Length rugosity (z_0)	0.001		
Rugosity class	0.32		
Wind profile	0.11		
Electrical loss	3%		
Wake effect loss	2%		
Unavailabilities	3%		
Others	1%		
Total	9%		

- Weibull distribution methodology is applied for the calculation of the wind velocity probability at measurement height and hub height of the wind turbine based on the Weibull parameters obtained. Therefore, the main results obtained are described:
  - Probability distribution of the wind velocity function p(v).
  - Wind potential exploited by the wind turbine in the case that its hub would be located at the measured height and the real wind potential exploited by the wind turbine at the real height where the hub would be located. The result is 342.79 MW/m^2 of wind

potential at the measured height and 344.96 MW/m^2 of wind potential at the hub height. In a complementary way, the gross and net wind power has been calculated, obtaining 4.72 MW gross power and 4,0 MW net power.

- Distribution of annual hours for each wind velocity or annual distribution of wind velocity, expressed in hours per year. It is calculated as the probability for each wind velocity multiplied by the annual hours, 8,760 hours.
- Density of electrical energy used by the wind turbine, expressed in MWh/m^2. It is calculated as the sum of the products of the probability by the corresponding power of the wind turbine, for each wind velocity, divided by the swept area of the wind turbine. A gross electric energy density of 3.02 MWh/m^2 and a net electric energy density of 2.75 MWh/m^2 is obtained.

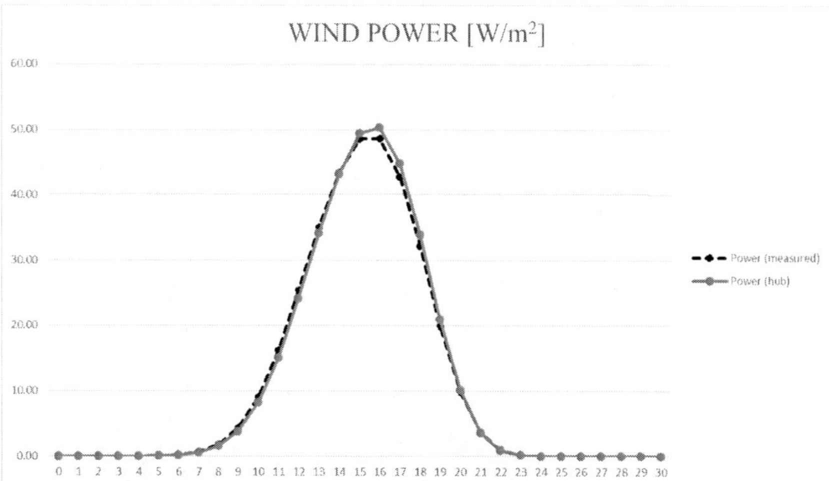

Figure 6. WTG Wind power generation distribution.

- Annual production of the wind turbine, gross and net, expressed in MWh/year. A gross production of 47.4 GWh/year and a net production of 37.6 GWh/year are calculated. The gross production is obtained as the sum of the annual production for each wind velocity. In a complementary way, a gross annual production of the installation of 1,240.6 GWh/year and a production of 1,128.9 GWh/year are calculated.

- Electrical efficiency of the wind turbine in each velocity class.
- Annual mean overall performance, load or capacity factor, equivalent hours and operating factor. They are calculated from the expressions described above. An annual mean overall efficiency of 0.14 is calculated. The capacity factor is 0.86. Finally, equivalent hours of 7,526 hours and an operating factor of 85.92% are calculated.

- Finally, Figure 6 shows the distribution of the wind power used by the wind turbine generator at both reference points, measurement and hub.

Final Considerations and Conclusion

- Statistical calculation methodology of the parameters of centralization and dispersion has been applied from database that collect the number of annual hours in which wind velocity is located within the configured classes.
- Specifically, as one of the solutions for the calculation of such parameters, is the application of Weibull distribution law to baseline data by means of least squares fit to the Weibull law. Scale and shape factor variables have been calculated.
- The parameters of mean annual velocity, median, variance, standard deviation, mean annual velocity of the cube of the velocities, effective velocity, wind potential and energy density of the site at the measuring height per unit area and wind turbine hub and the wind potential and energy density of the site have been calculated.
- For the further development of the methodology, a specific wind turbine model, Gamesa SG 5.0-132, has been proposed to determine the parameters of swept area, power curve and hub height.
- Parameters have been proposed to determine the characteristics of the terrain.
- Specific losses have been proposed, which we classify into electrical losses, wake losses, scheduled and unscheduled unavailability and others.
- Weibull distribution method has been applied, through the shape and scale factors to represent the Weibull velocity distribution curves and cumulative frequency distribution.

- From the power curve of the selected wind turbine, annual distribution of velocities calculated at both heights (measurement and hub) and the annual hourly wind distribution is calculated.
- Gross and net annual wind potential extracted from wind turbine is calculated based on the calculated annual distribution and wind turbine parameters.
- Annual gross and net production of the installation is obtained based on the proposed number of wind turbines.
- Finally, additional parameters of mean annual overall performance, load or capacity factor, equivalent hours and operating factor have been obtained.

This chapter is a preliminary tool of simple application with the aim of parameterizing and measuring the energy potentiality of a possible site area for wind power generation based on wind measurement data in that site area. These data allow to analyse not only the potentiality of the site, but also to analyse different wind turbine models in order to find the optimal one to obtain the highest possible annual energy from the site. The modification of the technical parameters of the wind turbine in the calculation methodology will result in the new annual production and new statistical and performance parameters calculated, with a new reference hub height as it will depend on the characteristics of the new model.

On the other hand, this methodology also allows the addition of new hourly or hourly mean data with origin in data collected every minor period of time. This modification will affect the calculation of the shape factor and scale parameters of the Weibull function by least squares adjustment, i.e., it will modify the velocity distribution (calculated relative frequencies) but it will not be necessary to modify the velocity classes, in other words, it will not affect the developed methodology.

The analysis of the results obtained is what makes it possible to make the right decisions regarding the monitoring of the activities under study for the operation of a wind site. This methodology makes it possible to calculate sufficient parameters that will determine the possibilities of carrying out more specific wind modelling studies, extending the time periods (with historical or new measurement series) and measurement frequency or even placing a wind turbine on a pilot scale in order to obtain a more representative series of data. This development of the methodology itself adds robustness to it.

As shown, this methodology and its results are valid as starting data to address the economic aspects of wind energy, i.e., the calculation of the economic feasibility for its implementation in the study area. In the economic study analysis, two types of projects are distinguished:

a) A large electricity production facility for export to the grid, such as a wind farm.
b) A small or medium-sized installation, autonomous on an island or assisted by the external grid, or for self-consumption.

First case, more common for studies that apply the methodology described, involves the determination of the Levelized Cost of Energy (LCOE), measured in €/MWh. Profitability analysis of the installation, taking into account investment costs, O&M costs, among other less relevant costs (financing, amortizations, among others). Lately, it has to be taking into account the possible annual income that the installation can generate from the electricity production operation, which depends on energy market activities.

References

[1] "MetLink - Royal Meteorological Society Atmosphere -," *MetLink - Royal Meteorological Society*. https://www.metlink.org/resource/atmosphere/ (accessed Apr. 04, 2022).
[2] Alison Nugent, D. D., A. Nugent, D. DeCou, S. Russell, and C. Karamperidou, *Atmospheric Processes and Phenomenon*. Accessed: Apr. 04, 2022. [Online]. Available: http://pressbooks-dev.oer.hawaii.edu/atmo/
[3] "Inicio | Idae." https://www.idae.es/home (accessed Apr. 04, 2022).
[4] "Centro de Investigaciones Energéticas, Medioambientales y Tecnológicas." ["Center for Energy, Environmental and Technological Research."] https://www.ciemat.es/ (accessed Apr. 04, 2022).
[5] "The European Wind Energy Association| EWEA." http://www.ewea.org// (accessed Apr. 04, 2022).
[6] "Onshore Wind Turbine SG 5.0-132 I Siemens Gamesa." https://www.siemensgamesa.com/en-int/products-and-services/onshore/wind-turbine-sg-5-0-132 (accessed Apr. 04, 2022).
[7] "Wind energy database." https://www.thewindpower.net/ (accessed Apr. 04, 2022).
[8] López, M. V. *Ingeniería de la Energía Eólica*. Marcombo, 2011.

Chapter 4

State Feedback Controller Design for Batteries Participating in Load Frequency Control in Power Networks with Wind Generation

Semaria Ruiz[1,*] and Daniel Icaza[2]
[1]Energetic Process Department, Universidad Nacional de Colombia, Medellín, Colombia
[2]Electrical Engineering Department, Universidad Católica de Cuenca, Cuenca, Ecuador

Abstract

In this chapter, we propose an optimal feedback control strategy to manage the power exchanged between a battery bank and the power system, considering that the battery bank participates in the frequency control loop of a power system with wind power penetration. It is motivated by the fact that the stochastic nature of renewable generation induces a variability in the frequency of power systems, which must be compensated by renewable sources themselves, by the conventional generators, or by backup storage systems. Specifically, we analyze the compensation of these frequency changes using backup batteries. We explore the potential of their participation in two Load Frequency Control services: the primary and the secondary regulation. In this regard, we develop the control design methodology for both services based on the spectral characterization of wind power generation. Hence, the controllers will regulate the charging/discharging processes of the batteries to reject as much as possible the frequency disturbances induced by the changes in generation and load and consider the batteries' capacity

*Corresponding Author's Email: seruizal@unal.edu.co.

In: Wind Turbines
Editors: David Borge-Diez and Daniel Icaza
ISBN: 978-1-68507-974-1
© 2022 Nova Science Publishers, Inc.

limit. Therefore, we analyze the power spectrum of wind generation measurements provided in a previous study to characterize the generation disturbances. Furthermore, the problem of tuning the state feedback control parameters is proposed as a nonlinear optimization, solved with the Sequential Quadratic Programming algorithm.

Keywords: anti-windup, batteries, feedback control, frequency control, nonlinear optimization, wind power

Introduction

The use of electrical energy plays a crucial role in the socio-economic development of the population (Leung and Meisen 2005). Hence, nowadays a growing energy demand exists in urban and rural areas around the world, which needs to be satisfied by increasing generation capacity. In 2012, 68% of the energy generated in the world came from the burning fossil fuels (natural gas and oil) (Momoh 2012). However, these fuels are the main cause of environmental pollution, generating climate change worldwide, and their non-renewable nature will make them insufficient to meet the growing energy demand. Therefore, it has become a priority to use renewable energies in the generation process of electric energy (Soshinskaya et al., 2014)- (Ustun, Ozansoy, and Zayegh, 2011).

However, the inclusion of renewable sources in conventional power networks brings some difficulties and changes due to the intermittency of their energy since the power generated by sources such as wind power or photovoltaic modules depends on the climatic conditions of the installation site. For this reason, appropriate control and management strategies must be proposed for systems that include these sources, which must compensate for the variations in generation through the efficient use of storage devices or other compensation devices (Ustun, Ozansoy, and Zayegh 2011).

Specifically, some of the most representative challenges that must be faced in scenarios of high penetration of renewable sources in conventional systems are the increase in network failures and the loss of system inertia which can have a significant impact on the frequency and voltage stability of the network, the reduction in energy exchange capacities is due to the increase in local load flows, the overloading of transmission lines in regular operation and in emergency conditions, and maintaining power balancing and reserve capacities (Bevrani and Hiyama 2011), (Malik 2015).

Regarding the strategies for addressing the first and last challenges mentioned above, there is a concept known as ancillary services, which are services different from conventional scheduled generation intended to maintain the frequency and voltage stability of the network and to maintain the power/generation balance. Moreover, according to (Brooks, 2002), battery storage systems can provide various ancillary services such as frequency regulation, spinning reserve services, supplemental reserve services, and voltage support. In particular, frequency regulation or Load Frequency Control (LFC) is a well-paid service since its remuneration scheme is established explicitly in the most relevant electricity world markets: ISO-NE, NYISO, ERCOT, and the European markets from France, Germany, Norway, and Sweden. This service is related to the ability of batteries or generators to change their output power. Furthermore, it is classified into primary, secondary, and tertiary frequency regulation according to the response times of power changes. The primary regulation operates within a time interval of 10 s since a frequency event happens, and their participants require to maintain the power delivery for 30 s. The secondary regulation is activated 30 seconds after the occurrence of a low-frequency event and requires maintaining power delivery for 30 min. The tertiary regulation is activated 5 minutes after the occurrence of an event, and participants must hold power delivery for 2 hours (Greenwood et al., 2017).

Furthermore, as indicated by the authors of (Greenwood et al., 2017; Jansson, 2019), based on these definitions and the fact that the batteries' response to power changes is approximately 2.5-10 times faster than conventional generations, it is identified that the batteries have a high potential to provide primary and secondary frequency regulation services and to reject disturbances with high bandwidth, such as power generated by renewable sources. Hence, in this study, we take advantage of this potential and propose an energy management system for a battery bank aiming at mitigating the disturbance introduced by wind sources penetration into a conventional power system.

The abovementioned attractive properties of batteries have been considered in studies such as the ones presented in (Lu, Liu, and Wu 1995; Aditya and Das 1999; Yan et al., 2018; Biroon, Pisu, and Schoenwald 2021), which include the participation of batteries in the LFC control loop for conventional and non-conventional power systems. These studies use control strategies such as heuristics (Lu, Liu, and Wu 1995), PI controllers (Aditya and Das 1999), state feedback $H\infty$ controllers (Yan et al., 2018), and optimal control (Biroon, Pisu, and Schoenwald 2021) to deal with charging-discharging control of battery storage systems participating in the LFC control loop. It should be noted that out of the available approaches, only the heuristic ones can directly handle the energy and power limits of the battery bank, which is a limiting key issue mentioned by the authors (Pati and Mishra 2020).

Hence, in order to jointly design a controller able to manage the charging-discharging power of the battery bank, according to the disturbances induced by the inclusion of wind power in conventional power systems and the energy-power limits of the bank; we propose to use an optimal state feedback control approach, following the same line than the research (Biroon, Pisu, and Schoenwald 2021), but also addressing the energy-power limits with the anti-windup technique, as a novelty. Since model-based feedback controllers fit the control requirements better than the PI controllers, the first ones allowed to include multiple-input multiple-output systems, which is the kind of system involved in the load frequency control loop. Furthermore, we propose a nonlinear optimization approach for tuning the state feedback frequency controller. This approach requires a previous definition of the desired location of the dominant closed-loop poles of the dynamic system, which is determined based on the characterization of the wind power disturbance frequency. Then, the optimization will provide the best values (which minimize the peak overshoot of the closed-loop system) for the following system elements: the location of the non-dominant poles, the eigenfrequency of the closed-loop system, the damping, the droop gain for conventional generators, the integral gain for the secondary frequency loop, and the gain of the anti-windup function.

The content of this chapter is divided into five sections as follows: the first section gives an overview of the frequency control concept, the frequency control applied to conventional power systems (composed of hydraulic and thermal generation), the frequency control under the inclusion of renewable sources (in this case wind power), and the frequency control under the inclusion of storage systems that participate in the frequency regulation task.

The second section presents technical details of the newly proposed methodology, starting with the description of the general control scheme used for frequency regulation. Then, we present the particularities of the state feedback control and a technique for calibrating its tuning parameters such that a desired frequency response of the closed-loop system is achieved. After that, we introduce the anti-windup concept and its integration with the state feedback control technique and develop the plant and controller models for our specific system. In this regard, we present the dynamic elements of our system, namely the conventional generators, the battery bank, and the renewable wind sources. Finally, we present the complete control scheme integrating the abovementioned elements and the anti-windup function limiting the battery bank power and energy content together with an algorithm summarizing the design of this controller. This design procedure focuses on tuning of the parameters of the controller and is formulated as a nonlinear optimization problem.

The third section introduces the power system standing for the case study used to verify the proposed control strategy. Attention is paid to the description of the system parameters and the wind sources' power spectrum considered in the case study.

The chapter is concluded by the "Results an analysis" section, where we describe the optimization procedure used to solve the problem of tuning the controller parameters. The solution obtained as a result of the parameter tuning optimization is presented together with the results of the simulation of the complete dynamic system in the Matlab Simulink toolbox.

Frequency Control

Frequency Control in Conventional Systems

The frequency in an electrical power system must lie within strict limits to guarantee the quality of the electric service provided to the final users, as the frequency deviations can cause malfunction of electrical equipment in industrial and residential loads. For example, some motors may be forced to rotate at different speeds, sensible electronic loads may be damaged, or clocks and automatisms that measure time depending on the power frequency may be delayed (Ledesma, 2008).

The frequency of an electrical system is closely related to the balance between generation and load. In a permanent operating regime, the network generators operate in synchronism. Under this condition, the accelerating torque applied by the turbines to the generators is equal to the electromagnetic torque, which tends to slow down the machine. If the system load increases, the electromagnetic torque increases, thus slowing down the generator and reducing the system frequency; and vice versa, if the system load decreases, the electromagnetic torque decreases, and the system frequency increases. That is why, to prevent frequency variations from causing negative impacts on the electrical system, most countries in the world have standards that regulate the limit values that the frequency can take.

Consequently, generators require a control system capable of responding to load and generation changes that occur in the system to keep its frequency within the permitted limits.

This control system is organized into primary, secondary, and tertiary levels.

The primary control is the fastest, operating in a time frame of between 2 and 20 seconds. It acts locally on each generator, considering the speed of rotation of the shaft. The inertia of the generators limits the speed of this control.

The secondary control operates within 20 seconds and 2 minutes. It acts within the scope of the control area, attending to the frequency and power exchange with neighboring areas.

The tertiary control operates in a time frame more significant than 10 minutes. It operates within an extensive electrical system, seeking load sharing and optimizing some chosen cost functions. Since most power systems worldwide do not have this third control loop, it will not be included in the present work.

The moments in which the frequency control loops act with a disturbance in the system are illustrated in Figure 1.

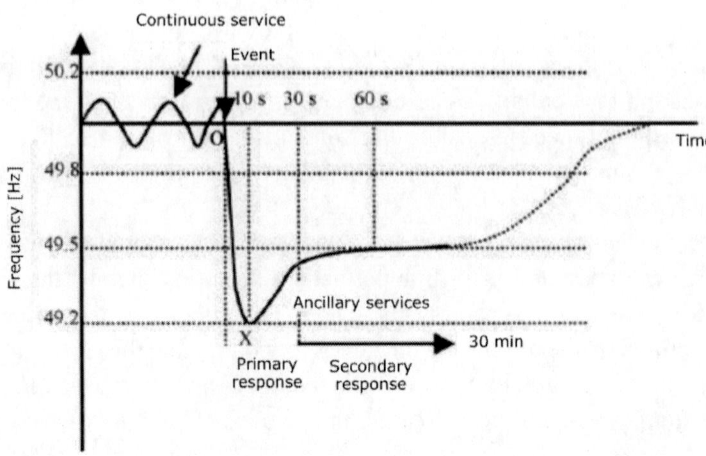

Figure 1. Schema of the frequency control system (Moore and Ekanayake 2009).

On the other hand, as is shown in Figure 1, the primary and secondary control loops are also called ancillary services; therefore, we also use this denomination to refer to the primary and secondary frequency regulation in the following work content. Moreover, we did not include the tertiary loop in the analysis, as we mentioned before. Hence, our proposed control strategy (deeply explained in the Methodology section) is intended to schedule the primary and secondary frequency control contributions for a power system composed of conventional generation units, a battery storage bank, and wind renewable generation.

Delving in details, as Figure 2 illustrates, in conventional power systems composed of fossil-fueled, hydraulic, or non-renewable energy generation, the primary and secondary loops of frequency control use Proportional and Proportional Integral (PI) controllers, respectively.

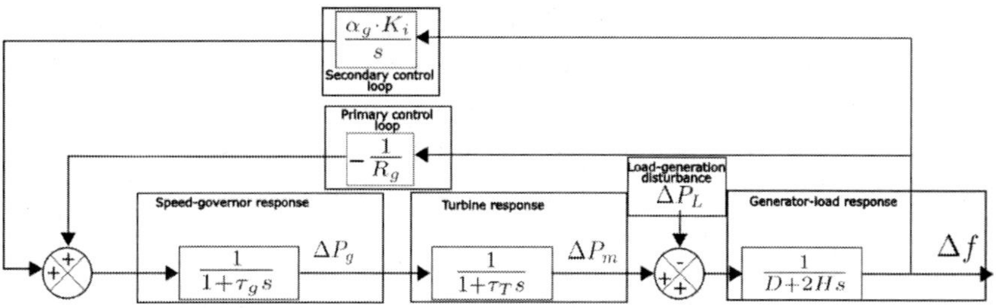

Figure 2. Model of the set: speed-governor, turbine, generator and load (Saadat, 1999).

Where K_i is the gain of the integral action and $\frac{1}{R_g}$ is the gain of the proportional action.

The representation of Figure 2 is an equivalent model of an isolated power system composed of conventional generation and loads sensible to frequency changes. It includes the main elements that must be modeled to represent the frequency response of a conventional generation unit. These elements are the turbine speed governor, which is responsible for sending a speed error signal to the turbine and used to modify the mechanical power. The turbine that is responsible for providing the mechanical power to the generation unit, the generator, which converts the mechanical power of the turbine into electrical power, and the system load, which may or may not be sensitive to frequency changes. In the case that it is sensitive, its sensitivity is modeled by the damping coefficient D. Also, we note that the scheme of Figure 2 considers one single lumped response for all generators and loads in the power system, which is represented in the block "Generator-load response." However, this lumped model is valid only for small-size power systems with conventional generation. Here we use this lumped model with the modifications set in the Methodology chapter, but we highlight that there exist more accurate models that can be used instead of the lumped such as the multi-area model.

Frequency Control Considering the Inclusion of Renewable Sources

The classic control scheme in Figure 2 has been modified in recent times to allow the inclusion of renewable sources, which affect the balance between load and generation of the power system due to the variability of the power generated by them. Since this generated power depends on stochastic climatological variables such as wind speed and solar irradiance, this fact has been demonstrated by the authors (Lalor, Mullane, and O'Malley 2005), who made a comparison between the variation of the frequency before the penetration of wind turbines in a conventional power system. They found that wind turbines generate frequency disturbances, which need to be compensated by adding surplus static reserve of energy in the power system. Then, with this new reserve, the power system could avoid instability in its frequency. In addition, the authors of the study (Yanagawa et al., 2001) have also found that frequency fluctuations occur when photovoltaic panels are integrated into conventional power systems, which could also cause instability problems.

For this reason, in the literature, many studies address the issue of the minimization of impact due to the inclusion of renewable sources on the frequency and power system stability. Here are some of these strategies:

The loads are included in the frequency control system in the studies (Short, Infield, and Freris 2007) and (Fabozzi, Thornhill, and Pal 2013). The authors proposed stabilizing the system through the dynamic control of load demand; in other words, some loads are switched off when a load-generation imbalance occurs to maintain the power network stability. Other studies, such as those presented in (Wei, Joos, and Abbey 2006; Kinjo et al., 2006), explore the inclusion of storage systems to maintain the frequency stability in the face of disturbances generated by the inclusion of wind turbines in the power network. In (Wei, Joos, and Abbey 2006), an energy storage system based on a filtering algorithm is proposed, that attenuates the frequencies at which fluctuations due to wind turbines occur; for variable speed wind turbines, which usually operate at maximum power. Furthermore, in the study (Kinjo et al., 2006), the authors propose an energy storage system composed of double-layer capacitors and power converters to compensate for the fluctuations generated by wind turbines in the power system. We have noted that in the present study, we have adopted the same idea as the studies (Wei, Joos, and Abbey, 2006; Kinjo et al., 2006), considering a large-scale implementation of Lithium-ion batteries to compensate for frequency changes since the economic feasibility of large-scale storage Lithium-ion systems has been improved in the recent years. According to the authors (Baumgarte, Glenk, and Rieger 2020), we are decreasing US$ 944 /kWh from 2010 to 2019 (Lithium-ion battery storage had a cost of US $156/kWh in 2019).

Finally, some methods other than the inclusion of storage systems, have been proposed in the literature to avoid frequency instability with the inclusion of renewable sources. This is the case of the study presented (Morren et al., 2006). The authors developed a method to control variable-speed wind turbines by making them emulate an inertial effect for the primary frequency control. This control proposes a variation of the turbine torque based on variations of the frequency and their change rate. We refer the reader to the paper (Ruiz, Patiño, and Espinosa, 2018) for more information about this topic.

Frequency Response and Energy Storage Systems

As indicated by the authors (Greenwood et al., 2017), there are several ways to represent or model the Lithium-ion battery energy storage systems for inclusion infrequency control. The most common models are the first-order transfer function proposed by the authors in reference (Arani & Mohamed, 2017). The equivalent circuit models (Swierczynski et al., 2014) and predictors of charge and degradation storage state (Swierczynski et al., 2014). Moreover, their participation in frequency control is considered in the primary loop, in the secondary loop, or in both. In our case, batteries are considered for participating in both control loops; and we use a first-order transfer function to represent the battery response, as Figure 3 illustrates. Additionally, we include a model to constrain the participation of batteries in frequency control, using the measurement of the SOC, which is deeply described in the Methodology section.

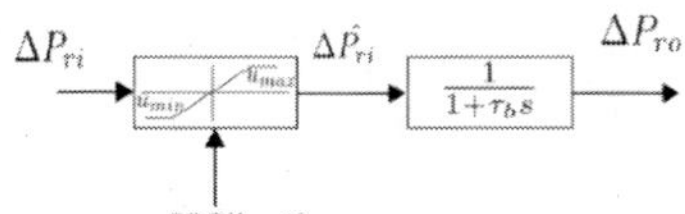

Figure 3. Block diagram of the battery model.

This SOC estimation model requires the previous definition of the battery bank capacity E_M. This variable must be calculated according to the percentage of renewable energy inclusion and the size of the conventional generation. In our case, we have calculated this value with the following criteria, established in (Greenwood et al., 2017): the size of the storage system will be set between 5%-10% of the total generation capacity.

Methodology

Control Scheme for the Provision of Frequency Regulation

First, we highlight our interest in designing a controller based on frequency deviation measurements that automatically schedules battery power according to these changes. Additionally, we use a state feedback control to perform this task, as indicated before. A general scheme of this controller is illustrated in Figure 4.

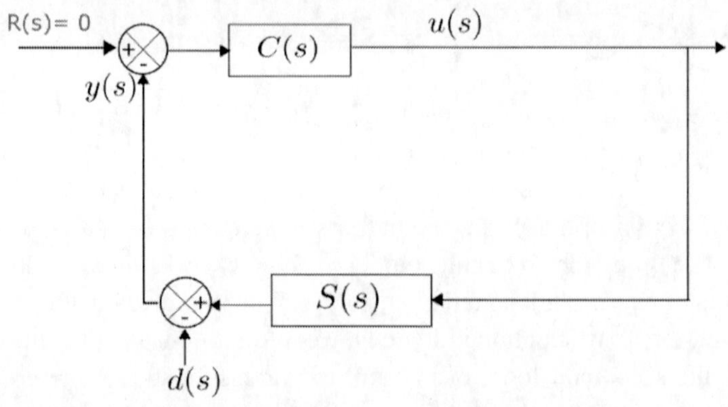

Figure 4. General state feedback control scheme.

Where $C(s)$ represents the controller transfer function, $S(s)$ is the plant transfer function, $u(s)$ is the control action (also called plant input variable), $y(s)$ is the output or controlled variable, $R(s)$ is the output reference value (set to 0 because this is the reference value for the frequency deviation), and $d(s)$ is the system disturbance.

State Feedback Control

As the authors indicated (Fadali and Visioli 2020), a state feedback controller uses the state-space model of a dynamic system to generate a gain matrix that guarantees the stability of the closed-loop system.

In a mathematical form, the state feedback control relies on representing a system in the way indicated by Equation (1).

$$\dot{x} = Ax + Bu \tag{1.a}$$

$$y = Cx \tag{1.b}$$

or its discrete version.

Where u is the vector of control variables, which have the general form of Equation (2):

$$u = K \cdot x \tag{2}$$

It is K the feedback matrix that needs to be defined so that the closed-loop system given by Equation (3) is stable.

$$\dot{x} = (A + BK)x \tag{3}$$

According to the author (Nathdatta, 2004), the continuous dynamic system (3) is stable if the following criteria are fulfilled: the pair (A,B) is stabilizable, or equivalently, there exists K such that the poles of the closed-loop system $(A + BK)$ are in the left semi-plane, i.e., their real part is lower than zero.

Furthermore, there exists a relationship between this closed-loop system stability and the closed-loop poles of the system, described in the following section.

Closed-Loop Poles

The closed-loop poles of a system are defined as the eigenvalues of the closed-loop system matrix $(A + BK)$ (Nathdatta, 2004); hence they are calculated as the roots of the characteristic Equation (4); which has n possible solutions for

m input variables, where n and m are related with to size of the system considering that $A \in \mathbb{R}^{n \times n}, B \in \mathbb{R}^{n \times m}, K \in \mathbb{R}^{m \times n}$.

$$\det\big(sI - (A + BK)\big) = 0 \tag{4}$$

And as we have indicated before, we have the condition that the real part of these roots must be lower than zero to guarantee the system's stability.

Additionally, it should be noted that depending on the values taken from these roots, the dynamic response of the closed-loop system will change. For example, for a system with two state variables (i.e., $n = 2$), the characteristic Equation will have the structure of a second-order polynomial, as the Equation (5) indicates (Fadali & Visioli, 2020):

$$s^2 + 2\zeta w_n s + w_n^2 = 0 \tag{5}$$

Or factoring this polynomial, we obtain Equation (6):

$$\left(s + \zeta w_n + j w_n \cdot \sqrt{1 - \zeta^2}\right) \cdot \left(s + \zeta w_n - j w_n \cdot \sqrt{1 - \zeta^2}\right) = 0 \tag{6}$$

where ζ is the over-shoot and w_n the natural eigenfrequency of the closed-loop system. Now, supposing that the settling time t_s and an overshoot ζ are specified; then, the natural eigenfrequency can be determined from the Equation $t_s = 4/(\zeta \cdot w_n)$, as also the desired characteristic polynomial (Boyd and Barratt 1991). Alternatively, the polynomial can also be defined by establishing a desired bandwidth for the controller BW and then determining the overshoot and the characteristic polynomial with the bandwidth relationship of Equation (7).

$$w_{BW} = \frac{4}{t_s \zeta} \sqrt{(1 - 2\zeta^2) + \sqrt{4\zeta^4 - 4\zeta^2 + 2}} \tag{7}$$

However, if the system states are higher than two (as we will show in later Sections, our system has six states), the characteristic polynomial will have more roots that need to be located. Here, we have adopted the criteria of defining $n - 2$ non-dominant poles with a magnitude between 2-3 times higher than the first two dominant poles calculated with previously defined criteria. Hence, the characteristic polynomial of our system will have the structure of Equation (8):

$$\left(s+\zeta w_n + jw_n\cdot\sqrt{1-\zeta^2}\right)\cdot\left(s+\zeta w_n - jw_n\cdot\sqrt{1-\zeta^2}\right)\cdot\left(s+g_1\cdot\left(\zeta w_n + jw_n\cdot\sqrt{1-\zeta^2}\right)\right)\cdot\left(s+g_1\cdot\left(\zeta w_n - jw_n\cdot\sqrt{1-\zeta^2}\right)\right)\cdot\left(s+g_2\cdot\left(\zeta w_n + jw_n\cdot\sqrt{1-\zeta^2}\right)\right)\cdot\left(s+g_2\cdot\left(\zeta w_n - jw_n\cdot\sqrt{1-\zeta^2}\right)\right)=0; \quad (8)$$

$2 \le g_1 \le 3;$

$2 \le g_2 \le 3;$

To summarize, we must determine the following controller parameters for the specific case of frequency regulation: i) the settling time t_r, which is the time required for the controller to reach a reference signal. It must be lower than the minimum response time required for the frequency regulation service set in the electric regulation codes. ii) The closed-loop characteristic of the controller, is associated with the location of the closed-loop system poles that guarantee a desired frequency response of the controller. In this case, a minimum required bandwidth exists, which must allow the rejection of the typical wind renewable energy disturbances and the typical load/generation variations in the system. And iii) the maximum-minimum energy and power of the batteries' bank.

Nonetheless, the closed-loop poles location procedure does not include a limitation for the input action, hence is not possible to address the constraints for energy and power of batteries by applying this procedure only. Thus, we propose to include a new element in the control loop to address this issue, called anti-windup.

State Feedback Control and Anti-Windup

The solution obtained for the feedback control law K, applying the closed-loop pole's location technique, does not consider constraints for the input or the state variables. Moreover, this is a critical issue to consider for the control of dynamic systems because the actuators and system outputs have to limit values; therefore, in a realistic control situation, the unlimited control actions are not available and are necessary to reflect this saturation process in the control modeling.

This is why the anti-windup schemes have been proposed, and apart from anti-windup, other techniques such as dual optimal control and model predictive control have been developed to overcome this issue. Nevertheless, here we focus only on the anti-windup schemes as they are the simplest among the constrained control techniques.

There are two anti-windup schemes: the constrained input and the constrained state.

As those presented in (Kothare et al., 1994), the state constrained ones require the implementation of Kalman filters for the state variables estimation. On the other hand, the input constrained schemes addressed in (Rojas & Goodwin, 2002) require the addition of a saturation block before applying the input to the plant and the inclusion of a new state variable in the closed-loop system model. We have used this last approach for our implementation since it is more easily adaptable to the problem of energy and power limits in batteries.

The general scheme of the anti-windup control with input constraints is illustrated in Figure 5. On it, it can be observed two new elements regarding the initial model presented in Figure 4, which enter the controller function $C(s)$:

An integral action regarding the signal error given by the block $\frac{K_i}{s} \cdot (y(s) - R(s))$, and the second integral action $\frac{K_c}{s} \cdot (u(s) - \hat{u}(s))$, related with the error between the unsaturated input $u(s)$ and the saturated input $\hat{u}(s)$.

On it, the anti-windup effect of the output constrained system is given by the term $\frac{K_c}{s} \cdot (u(s) - \hat{u}(s))$ which makes that the differences between the saturated and unsaturated inputs tend to be zero; hence, the system will be able to compensate the saturation output effect. The other term $\frac{K_i}{s} \cdot (y(s) - R(s))$ is included in the control scheme to avoid stationary state errors.

Now, defining two new state variables $\eta(s) = \frac{K_i}{s} \cdot (y(s) - R(s))$ and $v(s) = \frac{1}{s} \cdot \left(-K_c v + \left(u(s) - \hat{u}(s)\right)\right)$, from the diagram of Figure 5 we can obtain the dynamic equations for this new state variables as:

$$\dot{\eta} = K_i \cdot (y - R) \tag{9.a}$$

$$\dot{v} = -K_c \cdot v + (u - \hat{u}) \tag{9.b}$$

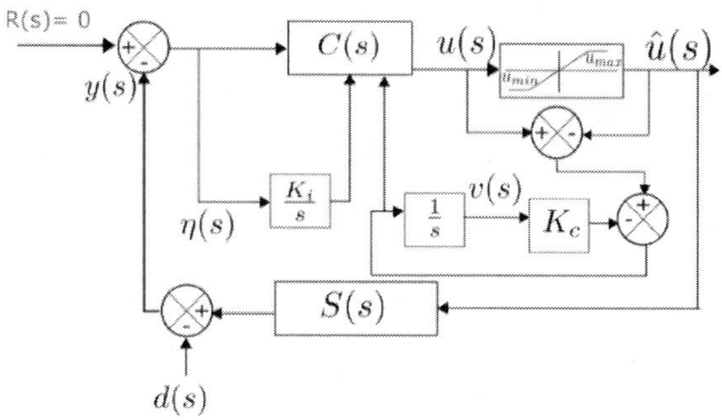

Figure 5. Proposed control scheme with anti-windup.

With this complete anti-windup control scheme in mind, we apply the previously defined concepts to our particular dynamic system in the next section.

Plant and Controller Modelling

For tuning the parameters as mentioned earlier, we must firstdeduce the plant and controller transfer functions. Furthermore, as we set previously, the controller proposed is a state feedback one. Meanwhile, the plant is modeled in three parts: a first-order system to represent the response of the battery bank to the frequency changes (Arani & Mohamed, 2017) shown in Figure 3 and depicted in Equation (10), two first-order systems to represent the response of the conventional generators to the frequency changes, and the power system frequency response indicated in equation (12), taken from reference (Saadat, 1999) and shown in Figure 2.

Furthermore, other states and inputs are added to the plant model to include the anti-windup dynamics, described by Equations (9) and (15).

Lest first describe the battery model, which is deduced from Figure 3, omitting the saturation block and making $\Delta P_{ri} = \Delta P_{ref,b}$:

$$\Delta P_{ro} = \frac{1}{1+\tau_b\, s} (\Delta P_{ref,b}) \tag{10}$$

where τ_b is the time constant of the battery response given by the manufacturer, $\Delta P_{ref,b}$ is the power required for the batteries' bank to provide frequency regulation and ΔP_{ro} is the power exchanged between the batteries' bank and the power system (i.e., the contribution of the batteries to the load frequency control).

On the other hand, the representation of speed-governor and turbine responses are given in Equations (11). These equations are obtained from Figure 2, replacing the PI control function in the secondary control loop with a generic error signal $\Delta P_{ref,g}$.

$$\Delta P_m = \frac{1}{1+\tau_T\, s}\,\Delta P_g \tag{11.a}$$

$$\Delta P_g = \frac{1}{1+\tau_g\, s}\left(-\frac{1}{R_g}\Delta f + \Delta P_{ref,g}\right) \tag{11.b}$$

where τ_T is the time constant of the conventional generation turbines, τ_g is the time constant of the conventional generation units, ΔP_m is the change in the turbine's mechanical power, . ΔP_g is the change in the power supplied by the conventional generation units, R_g is the droop gain of the conventional generators related to the primary frequency response, and $\Delta P_{ref,g}$ is the contribution of conventional generators to the secondary loop of frequency regulation.

Finally, the dynamic response of the generation and loads in the power system to the frequency changes is given by Equation (11).

$$\Delta f = \frac{1}{D+2\,H\,s}\left(\Delta P_m + \Delta P_{ro} - \Delta\, P_L\right) \tag{12}$$

where Δf indicates the frequency deviations, D is the damping coefficient of the power system, H is the equivalent system inertia, and $\Delta\, P_L$ represents the load-generation disturbances; particularly, in this case, the variable $\Delta\, P_L$ is related to the changes generated by the inclusion of wind sources. In the next section, we describe in detail the modeling of the wind generation through this variable ΔP_L.

Wind Generation Model

Wind turbines are divided into two broad categories: constant and variable speed. Variable-speed wind turbines tend to be more common today because they can seek the point of maximum power extraction by not maintaining their speed. Therefore, in this work, we consider the inclusion of variable speed wind turbines in a conventional system.

Various models for variable speed turbines are available in the literature; below, we will describe the most used.

- The mechanical model (Thomsen, 2006) describes the aerodynamic equations of the wind turbine, which relate to the wind speed, the angular speed of the turbine, and the pitch angle.
- The circuital model (Ramtharan, Ekanayake, and Jenkins 2007) contains the dynamic equations for the induction generator of the wind turbine and the converters that act as an interface between the wind turbine and the power network. The current and voltage in the rotor of the induction generator are variables of interest in this model, which are controlled to regulate the generator torque.
- The frequency response model (Moore and Ekanayake 2009) includes the response of the wind turbines to frequency changes, considering they participate actively in the frequency control. Two assumptions to perform this are made: the wind turbines are operated under the maximum power point, and the wind turbines are equipped with a droop control unit that emulates the governor's behavior in conventional generation units.

Here, we do not consider wind turbines actively participating in the frequency control. Then, the implicit model we used is mechanical, representing the power changes due to the wind speed variations. Consequently, we assume that the wind turbines are operated at their maximum power point. In summary, we consider that the power spectrum illustrated in Figure 11 comes from variable speed wind turbines operated at their maximum power point.

Therefore, below we will describe in detail the mechanical model.

Let's calculate the mechanical power of the wind turbine $P_{m,w}$, and assume that the electric power generated P_L is an efficiency coefficient η_w multiplied by the mechanical power P_m. Hence, we have the next relationship $P_L = \eta_w \cdot P_{m,w}$.

Now, the mechanical power of a wind turbine is calculated with the expression of Equation (13):

$$P_r = \frac{1}{2}\rho\pi R^2 v^3 c_p(\lambda, \theta) \quad (13)$$

where ρ is the air density, R is the blade radius, v is the effective wind speed, c_p is the efficiency coefficient, which is a function of the angle of rotation of the blades θ, the speed of the blade's tip (also called rotational speed) ω_r and the wind speed v. This turbine efficiency coefficient c_p can be obtained through numerical calculations or manufacturer's tables. In this case, the coefficient is calculated from a mathematical expression extracted from (Slootweg, Polinder, and Kling 2001) and depicted in Equation (14).

$$c_p(\lambda, \theta) = 0.22\left(\frac{116}{\lambda_t} - 0.6\theta - 5\right)\exp\left(-\frac{12.5}{\lambda_t}\right) \tag{14}$$

where λ_t is a coefficient that complies with the expression of Equation (15).

$$\frac{1}{\lambda_t} = \frac{1}{\lambda^{-1}+0.12\theta} - \frac{0.035}{(1.5\theta)^3+1} \tag{15}$$

and λ is another coefficient (also called wind speed-rotational speed ratio) which is calculated with Equation (16).

$$\lambda = \frac{v}{R\cdot\omega_r} \tag{16}$$

On the other hand, we can also obtain the mechanical torque of the wind turbine, using the previous expressions, as shown in Equation (17):

$$T_{m,w} = \frac{P_{m,w}}{\omega_r} \tag{17}$$

Then, replacing Equations (14)-(17) in (13), we can obtain the expression of the mechanical torque:

$$T_{m,w} = \frac{\rho\pi R^2 v^3}{2\cdot\omega_r}\cdot 0.22\left(\frac{116}{\frac{1}{\frac{1}{\frac{r\cdot\omega_r}{v}+0.12\theta}-\frac{0.035}{(1.5\theta)^3+1}}} - 0.6\theta - 5\right)\cdot \exp\left(-\frac{12.5}{\frac{1}{\frac{1}{\frac{r\cdot\omega_r}{v}+0.12\theta}-\frac{0.035}{(1.5\theta)^3+1}}}\right) \tag{18}$$

We can appreciate from Equation (18) that the mechanical torque of the wind turbine depends on the rotational speed ω_r, the wind speed v, and the pitch θ. Several methods control the variable speed wind turbines; some focus on pitch control, others on the angular speed control, and others more focused on the power control.

If we assume that the pitch angle θ is fixed at a certain value, we can find the curves related to the torque, the rotational speed, and the wind speed of the turbine. These curves are illustrated in Figure 6. They were obtained for a wind turbine of 2 MW (Ruiz, Patiño, and Espinosa, 2018).

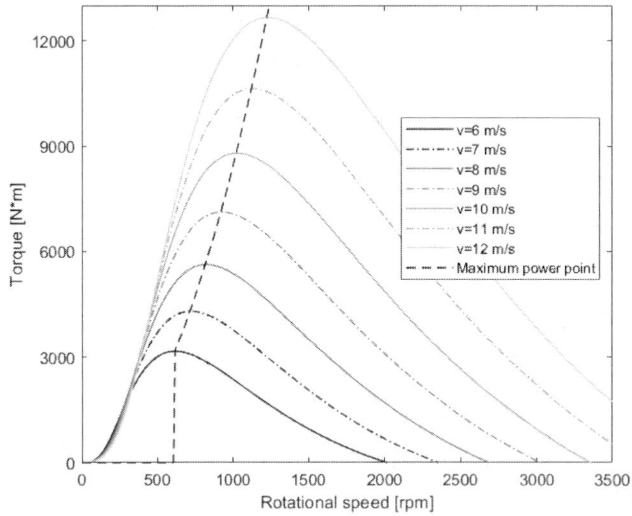

Figure 6. Torque-rotational speed curves of a speed variable wind turbine.

According to the curves of Figure 6, we assume that the pitch angle is fixed. The wind turbines have an integrated controller that brings the rotational speed and the torque of the turbine to the point at which it is possible to extract the maximum power for a certain wind speed.

Hence, the wind turbine rotational speed ω_r and the mechanic torque $T_{m,w}$ will be determined as the optimal points of the function (18). Consequently, the obtained mechanical power $P_{m,w}$, and the electrical power generated by the wind turbine P_L will vary with the wind speed.

Complete Control Scheme

By joining the models mentioned above, we can obtain the system illustrated in Figure 7, whose state-space representation is described in Equation (19), which considers that batteries and conventional generators will participate in the frequency regulation control loop.

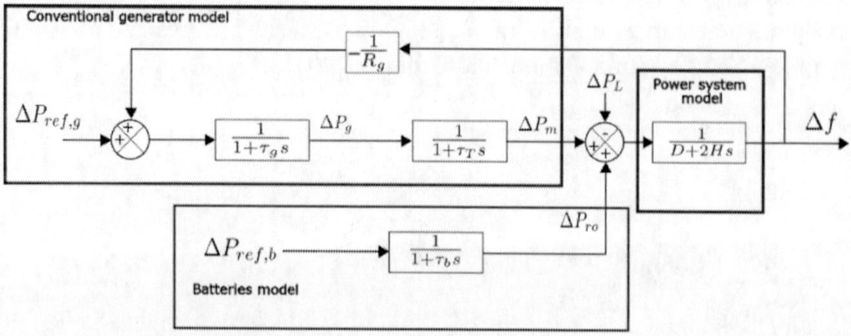

Figure 7. Block diagram of the complete system.

$$\begin{bmatrix} \dot{\Delta f}(t) \\ \Delta\dot{P}_g(t) \\ \Delta\dot{P}_m(t) \\ \Delta\dot{P}_{ro}(t) \end{bmatrix} = \begin{bmatrix} -\frac{D}{2H} & 0 & \frac{1}{2H} & \frac{1}{2H} \\ -\frac{1}{\tau_g R_g} & -\frac{1}{\tau_g} & 0 & 0 \\ 0 & \frac{1}{\tau_T} & -\frac{1}{\tau_T} & 0 \\ 0 & 0 & 0 & -\frac{1}{\tau_b} \end{bmatrix} \begin{bmatrix} \Delta f(t) \\ \Delta P_g(t) \\ \Delta P_m(t) \\ \Delta P_{ro}(t) \end{bmatrix} +$$

$$\begin{bmatrix} 0 & 0 \\ \frac{1}{\tau_g} & 0 \\ 0 & 0 \\ 0 & \frac{1}{\tau_b} \end{bmatrix} \begin{bmatrix} \Delta P_{ref,g}(t) \\ \Delta P_{ref,b}(t) \end{bmatrix} + \begin{bmatrix} -\frac{1}{2H} \\ 0 \\ 0 \\ 0 \end{bmatrix} \Delta P_L(t) \tag{19}$$

where the states x are Δf the frequency deviations, $\Delta P_g(t)$ the power changes in the conventional generation units, $\Delta P_m(t)$ the changes in the mechanical power delivered by the conventional units, and $\Delta P_{ro}(t)$ the change in the power delivered/stored in the battery bank.

The control inputs u are the contribution of the conventional generators to the frequency regulation $\Delta P_{ref,g}$, and the contribution of the batteries to the frequency regulation $\Delta P_{ref,b}$.

The disturbance d is ΔP_L which models the changes in generation introduced by the wind turbines.

And the output variable y is the frequency deviation Δf.

Moreover, the model of Equation (19) can be represented in compact matrix form as Equation (20) indicates:

$$\dot{x} = A_s \cdot x + B_s \cdot u + D_s \cdot d \tag{20}$$

However, this model does not include the constraint for the energy and power of the batteries' bank since the state variable related to the power exchanges by the batteries ΔP_{ro} is neither constrained in Equation (19). Thus, we propose to include a new element in the control loop to address this issue, called anti-windup.

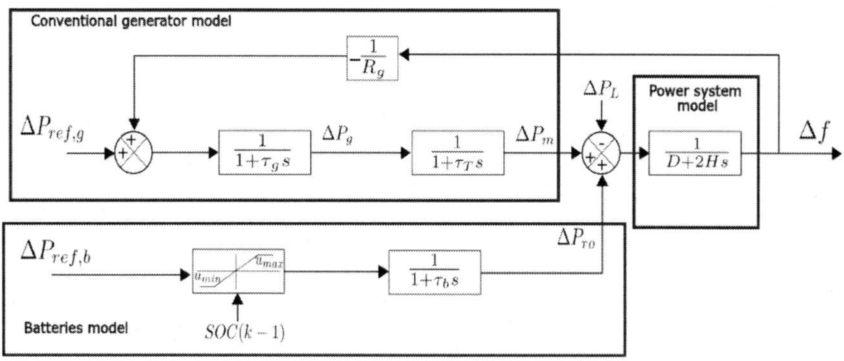

Figure 8. Block diagram of frequency control with anti-windup.

The state-space representation for the dynamic system, including the controller with the anti-windup element, is depicted in Equation (21). This control model, illustrated in Figure 5 with the additional elements indicated in Equations (9), was constructed based on the approaches proposed in the references (Kothare et al., 1994) (Dai et al., 2007).

In Equation (21), the control inputs are the variables $\left[\Delta P_{ref,b}\ \Delta P_{ref,g}\right]^T$, and the variable $\Delta\widehat{P_{ref,b}}$ is the saturated value of the input variable $\Delta P_{ref,b}$. This saturated value is calculated based on the energy and power limits of the battery with Equation (22):

$$\begin{bmatrix} \dot{\Delta f}(t) \\ \dot{\eta}(t) \\ \Delta\dot{P}_g(t) \\ \Delta\dot{P}_m(t) \\ \Delta\dot{P}_{ro}(t) \\ \dot{v}(t) \end{bmatrix} = \begin{bmatrix} -\frac{D}{2H} & 0 & 0 & \frac{1}{2H} & \frac{1}{2H} & 0 \\ K_i & 0 & 0 & 0 & 0 & 0 \\ -\frac{1}{\tau_g R_g} & 0 & -\frac{1}{\tau_g} & 0 & 0 & 0 \\ 0 & 0 & \frac{1}{\tau_T} & -\frac{1}{\tau_T} & 0 & 0 \\ 0 & 0 & 0 & 0 & -\frac{1}{\tau_b} & 0 \\ 0 & 0 & 0 & 0 & 0 & -K_c \end{bmatrix} \begin{bmatrix} \Delta f(t) \\ \eta(t) \\ \Delta P_g(t) \\ \Delta P_m(t) \\ \Delta P_{ro}(t) \\ v(t) \end{bmatrix} +$$

$$\begin{bmatrix} 0 & 0 \\ \frac{1}{\tau_g} & 0 \\ 0 & 0 \\ 0 & \frac{1}{\tau_b} \\ 0 & -1 \end{bmatrix} \begin{bmatrix} \Delta P_{ref,g}(t) \\ \Delta \widehat{P_{ref,b}}(t) \end{bmatrix} \begin{bmatrix} 0 \\ 0 \\ 0 \\ 0 \\ 0 \\ 1 \end{bmatrix} \begin{bmatrix} \Delta P_{ref,g}(t) \\ \Delta P_{ref,b}(t) \end{bmatrix} + \begin{bmatrix} -\frac{1}{2H} \\ 0 \\ 0 \\ 0 \\ 0 \\ 1 \end{bmatrix} \Delta P_L(t) \tag{21}$$

$$\Delta \widehat{P_{ref,b}} = \varphi(\Delta P_{ref,b}) \tag{22}$$

where $\varphi(\cdot)$ is a saturation function with variable limits that accomplish with Equation (23).

$$\varphi(x) = \begin{Bmatrix} u_{min} \; if \; x \le u_{min}; where \; u_{min} = max(P_{min}, \Delta P_{min}) \\ x \\ u_{max} \; if \; x \ge u_{max}; where \; u_{max} = min(P_{max}, \Delta P_{max}) \end{Bmatrix} \tag{23}$$

Here, the variable u_{min} indicates the lower bound for the required change in battery power $\Delta P_{ref,b}$, the variable u_{max} indicates its upper bound, P_{min} is the minimum allowed power exchange for the battery bank, P_{max} is the maximum allowed power exchange for the battery bank, $SOC(k)$ is the state of charge of the battery bank at the discrete-time step k, SOC_{min} is the minimum allowed depth of discharge for the battery bank, t_d is the discretization time for the estimation of the state of charge of the battery bank, S_{base} is the base power taken into account for calculating the system parameters in p.u. units, and ΔP_{min}, ΔP_{max} are the limits on the power changes related to the energy limits given by Equation (24).

$$\Delta P_{min} = \left(\frac{SOC_{min} - SOC(k-1)}{t_d \cdot S_{base}}\right) E_{max} \cdot 5 \cdot \tau_d \tag{24.a}$$

$$\Delta P_{max} = \left(\frac{SOC_{max} - SOC(k-1)}{t_d \cdot S_{base}}\right) E_{max} \cdot 5 \cdot \tau_d \qquad (24.b)$$

Next, we deeply explain the limits indicated in Equation (24): Let's consider first the expression to calculate the state of charge of the battery bank SOC for time step k, as Equation (25) indicates:

$$SOC(k) = SOC(k-1) + \frac{P_{bat}(k) \cdot t_d \cdot S_{base}}{E_{max}} \qquad (25)$$

where $\Delta P_{bat}(k)$ is a sample at the time k of the continuous variable representing the power delivered or stored by the battery bank. And, considering that battery reaches the reference value of the power output at $5 \cdot \tau_b$, we can approximate by a linear function, the relationship between ΔP_{ro} and $\Delta P_{ref,b}$ as: $\Delta P_{ro} = P_{bat} = \frac{\Delta P_{ref,b}}{5 \cdot \tau_b}$; to obtain a saturation function depending on the input variables. Then, the SOC is calculated with Equation (26).

$$SOC(k) = SOC(k-1) + \frac{\Delta P_{ref,b}(k) \cdot t_d \cdot S_{base}}{5 \cdot \tau_d \cdot E_{max}} \qquad (26)$$

Moreover, the state of charge of the battery will be limited between a minimum value SOC_{min}, and a maximum value SOC_{max}, according to the manufacturer's recommendations, to comply with the physical storage limits and to avoid battery wear. Therefore, the following inequalities must be met:

$$SOC(k) \geq SOC_{min} \qquad (27.a)$$

$$SOC(k) \leq SOC_{max} \qquad (27.b)$$

Then, replacing (26) in (27) we obtain the limiting values of the battery bank power given by Equation (28):

$$P_{bat}(k) \geq \left(SOC_{min} - SOC(k-1)\right) * \frac{5 \cdot \tau_d E_{max}}{t_d \cdot S_{base}} = \Delta P_{min} \qquad (28.a)$$

$$P_{bat}(k) \leq \left(SOC_{max} - SOC(k-1)\right) * \frac{5 \cdot \tau_d E_{max}}{t_d \cdot S_{base}} = \Delta P_{max} \qquad (28.b)$$

These limits are also illustrated in Figure 9.

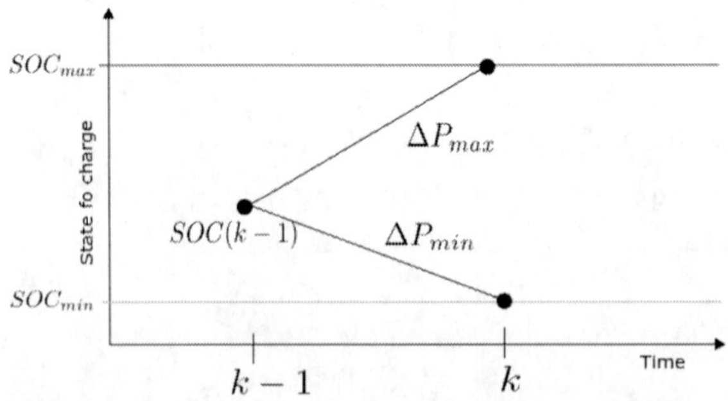

Figure 9. Illustration of energy limits for the battery bank.

Furthermore, we note that the saturation function $\varphi(\cdot)$ requires the value of the battery bank state of charge at a previous time step $SOC(k-1)$; thus, a sensor or an estimator of the state of charge for the battery bank must be implemented in order to make possible the system control. We have included an estimator that uses Equation (26) in our case.

On the other hand, the state variables of the Equation (21) are Δf the frequency deviations, η summation in time of the frequency error, $\Delta P_g(t)$ the power changes in the conventional generation units, $\Delta P_m(t)$ the changes in mechanical power delivered by conventional units, $\Delta P_{ro}(t)$ the change in the power delivered/stored in the battery bank, $v(t)$ summation in time of the difference between the saturated and unsaturated inputs $\Delta P_{ref,b}(t) - \Delta\widehat{P_{ref,b}}(t)$; which is multiplied by the weight factor B.

Moreover, Equations (21) can be alternatively written in a more general form, as in Equation (29):

$$\dot{x}(t) = A \cdot x(t) + B \cdot \hat{u}(t) + E \cdot u(t) + D \cdot d(t) \tag{29}$$

where $x = [\Delta f \quad \eta \quad \Delta P_g \quad \Delta P_m \quad \Delta P_{ro} \quad v]^T$ are the states of the system, $u = [\Delta P_{ref,g}\ \Delta P_{ref,b}]$ is the input to the system, $\hat{u} = [\Delta P_{ref,g}\ \Delta\widehat{P_{ref,b}}]$ is the saturated input, and $d = [\Delta P_L]$ is the disturbance due to wind renewable generation.

We should note that the described model does not consider saturation for the conventional generation required power $\Delta P_{ref,g}$: However, this phenomenon can be easily integrated with the proposed model.

Finally, we illustrate the complete diagram of the proposed control, including all the described variables in Figure 10.

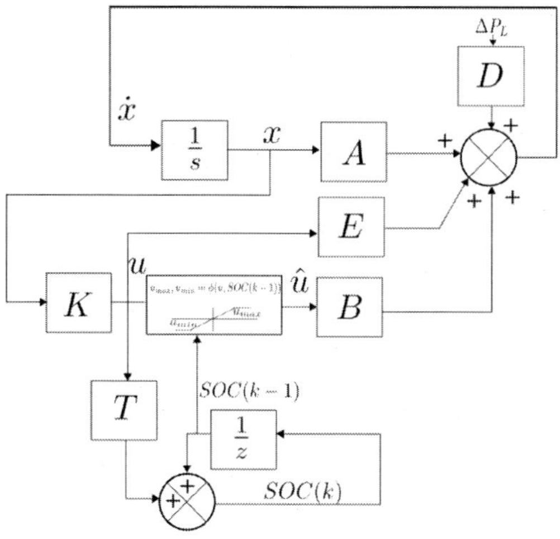

Figure 10. Scheme of the complete system state-space representation.

Where T is a matrix that selects only the input related to the battery's contribution and accomplishes the expression of Equation (30):

$$T = \left[0 \quad t_d \cdot \frac{S_{base}}{5 \cdot \tau_d \cdot E_{max}}\right] \tag{30}$$

Control Parameters Tuning

Summarizing, we proposed the following algorithm to tune the parameters of the state-feedback controller with anti-windup:

a) Establish a desired settling time $t_{r,des}$ according to the requirements of the frequency control provider and the desired bandwidth for the controller BW_{des}, based on the renewable source's variability characterization.
b) Determine the relationships between the setting time t_r, the bandwidth BW, the damping ration ζ, and the natural eigenfrequency w_n for the dominant second-order system behavior with Equation (7).
c) Determine the relationships between dominant and non-dominant poles with Equation (8).
d) Put together the above criteria to determine the damping ratio ζ and the natural eigenfrequency w_n, of the two dominant poles'

characteristic Equation, and the location of the other system poles. Taking into account that the closed-loop system will have six poles given by the characteristic Equation (5). Furthermore, considering the control parameters, K_c and K_i need to be determined.

Where the four non-dominant system poles are between 2–3 times faster than the real part of dominant poles (Massachusetts Institute of Technology - MIT 2010), that is why they are multiplied by the constants g_1 and g_2.

This step is done by solving the subsequent nonlinear optimization that minimizes the peak overshoot of the closed-loop system:

$$\min_{\zeta, w_n, g_1, g_2, K_i, K_c, R_g} \left(e^{-\zeta \cdot w_n \cdot \left(\frac{\pi}{w_n \sqrt{1-\zeta^2}} \right)} \right) \tag{31.a}$$

$Subject\ to$:

$$g_1, g_2 \geq 1; \tag{31.b}$$

$$g_1, g_2 \leq 3; \tag{31.c}$$

$$g_1 \cdot \zeta \cdot w_n \geq \zeta \cdot w_n; \tag{31.d}$$

$$g_2 \cdot \zeta \cdot w_n \geq \zeta \cdot w_n; \tag{31.e}$$

$$BW \geq BW_{des}; \tag{31.f}$$

$$t_r \leq t_{r,des}; \tag{31.g}$$

$$w_n \geq 0;\ \zeta \geq 0;\ \zeta \leq 1; \tag{31.h}$$

where the cost function (31.a) corresponds to the expression given by (Massachusetts Institute of Technology - MIT, 2010) to calculate the overshoot peak of the closed-loop system. The constraints (31.b) and (31.c) correspond to the range previously established for g_1 and g_2, which must lie between the interval [1,3]. The constraints (31.d) and (31.e) correspond to the rule established for the non-dominant poles of the closed-loop response, whose real part must be negative with a magnitude greater than the magnitude of the real part of the dominant poles. The constraint (31.f) is related to the accomplishment of the minimum required bandwidth, and the constraint (31.g) is related to the accomplishment of the minimum required time response.

Additionally, we want to point out that the decision variables of optimization (31) are the parameters related to the primary frequency control loop for conventional generators (R_g), the secondary frequency control loop (K_i), the anti-windup loop (K_c), and the poles position (w_n, ζ, g_1, g_2).

e) Calculate the position of the closed-loop system poles by solving the nonlinear characteristic Equation (4).

Study Case

The power system taken to apply the proposed frequency control strategy is a single area hybrid power system composed of conventional generation units, renewable wind sources, and a battery storage bank, as Figure 11 depicts.

The parameters of the system are shown in Table 1.

We should note that we have omitted the value of the load, as our analysis focuses on the transient response of the system, and we assume that load has a constant power demanded under the stable state regime. Furthermore, it is considered that the initial values for power generators met the active power balance condition; however, reactive powers are not set well to satisfice an optimal voltage condition in all system nodes.

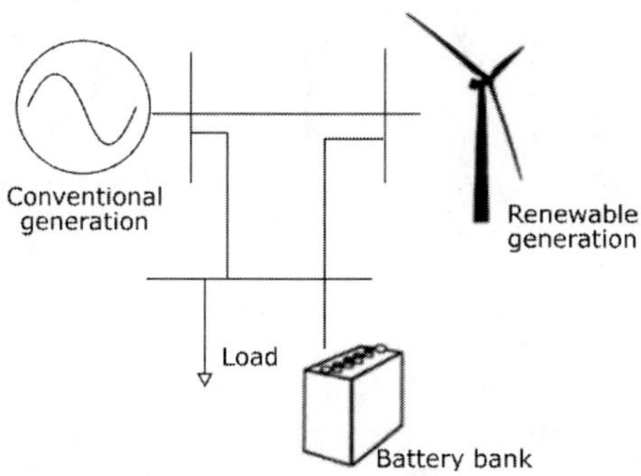

Figure 11. Illustration of the study case system.

Table 1. Study case system parameters

Parameter	Symbol	Value
Conventional generator inertia	H	3.01 [s]
Load sensitivity coefficient	D	0.8
Time delay for batteries response	τ_b	0.08 [s]
Time delay for governor	τ_g	0.2 [s]
Time delay for turbine	τ_T	0.3 [s]
Base power for calculations	S_{base}	100 [MVA]
Nominal frequency of the system	-	60 [Hz]
Maximum energy of the battery bank	E_{max}	11 [MWh]
Maximum power of the battery bank	P_{max}	3.67 [MW]
Minimum power of the battery bank	P_{min}	0 [MW]
Discretization time for the SOC model	t_d	0.1 [s]
Initial SOC	$SOC(0)$	0.5
Maximum SOC	SOC_{max}	1
Minimum SOC	SOC_{min}	0.2
Desired settling time	$t_{r,des}$	1 [s]
Desired bandwidth	BW_{des}	6.2832 rad [s]

Moreover, we have set the desired settling time and bandwidth of Table 1 based on the power spectrum of wind generation disturbances characterized in the reference (Bossuyt et al., 2016), which is described in detail in the following section.

Power Spectrum of Renewable Sources

Any time-varying signal can be expressed in frequency, which gives an idea of the periodic components of the signal. Moreover, the magnitude of each peak of the periodic component can be determined by the total energy of the signal, known as the obtaining of power spectral density.

On its side, if the inclusion of renewable sources is considered, the generation, besides the load, will be a source of disturbance in the system, as renewable generation, mainly wind power, presents fast changes. Authors in reference (Bossuyt et al., 2016) have characterized these changes of generation in the frequency domain and found that these changes are between a bandwidth of 0.05 - 60 Hz, approximately.

The power spectral density of the wind generator characterized in the study above is presented in Figure 12.

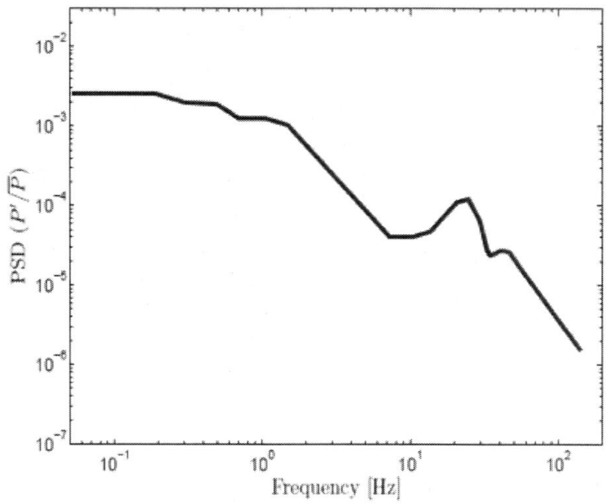

Figure 12. Frequency spectrum of renewable generation variations. Adapted from (Bossuyt et al., 2016).

Therefore, to emulate a signal with this spectrum, we have generated a random signal with a periodicity of 0.2 [s] and a maximum amplitude of 0.008 [p.u.], which is shown in Figure 13.

Results and Analysis

Solving the optimization indicated in Equation (31) with the previously described study case parameters, we obtain the results described below, Table 2.

The solution to the optimization problem was performed using the Sequential Quadratic Programming (SQP) algorithm of the Matlab optimization tool "fmincon."

The SQP algorithm is one of the most effective methods for solving nonlinear constrained optimization problems, as those presented in Equation (31). This method makes a second-order Taylor expansion of the cost function and a first-order approximation for the constraints, as follows.

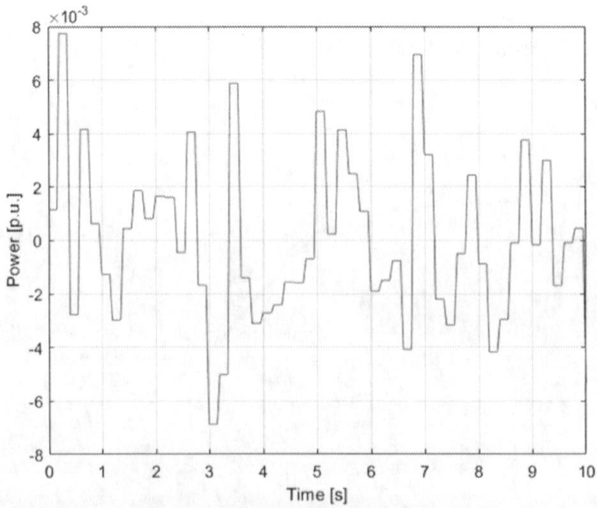

Figure 13. Generated wind power disturbance.

Table 2. Case study results

Result	Value
w_n	7.7994 [rad/s]
ζ	0.9727
g_1	2.3675
g_2	2.7766
K_i	-4.9662
K_c	16.9834
R_g	0.7721 [p.u.]

Suppose that we have the next nonlinear optimization problem.

$$\min_{x} \ f(x) \tag{32.a}$$

$Subject\ to$:

$$g_i(x) \leq 0, i \in \{1, \dots, m\} \tag{32.b}$$

$$h_j(x) = 0, j \in \{1, \dots, n\} \tag{32.c}$$

where $f(x)$ is the nonlinear cost function, and $g(x), h(x)$ are the nonlinear inequality and equality constraints respectively.

Then, considering the point x_k, is possible to make a Taylor expansion of f, g, and h around this point. Thus, let's suppose that we make a second-order Taylor expansion for f and a first-order expansion for g and h; the problem of Equation (32) will be transformed into the following one:

$$\min_{x} \hat{f}(x) = f(x_k) + \nabla f(x_k)(x - x_k) + \frac{1}{2}(x - x_k)^T \nabla^2 f(x_k)(x - x_k) \quad (33.a)$$

Subject to:

$$\widehat{g_i}(x) = g_i(x_k) + \nabla g_i(x_k)(x - x_k) \leq 0, i \in \{1, \dots, m\} \quad (33.b)$$

$$\widehat{h_j}(x) = h_j(x_k) + \nabla h_j(x_k)(x - x_k) = 0, j \in \{1, \dots, n\} \quad (33.c)$$

Now, the problem of Equation (33) can be solved since it is a quadratic problem with an existing global minimum in the feasible region that fulfils the Karush Kuhn Tucker (KKT) conditions within a defined confidence level.

These conditions are indicated in Equation (34):

$$\nabla \hat{f}(x) + \sum_{i=1}^{m} \lambda_i \nabla \widehat{g_i}(x) + \sum_{i=1}^{m} \gamma_j \nabla \widehat{h_j}(x) = 0 \quad (34.a)$$

$$\hat{h}_j(x) = 0 \quad (34.b)$$

$$\hat{g}_i(x) \leq 0 \quad (34.c)$$

$$\lambda_i(x) \geq 0 \quad (34.d)$$

were L is the Lagrangian of the optimization problem (33), and λ, γ. are the Lagrange multipliers related to the inequality and equality constraints respectively.

The obtained control matrix K by applying the proposed method with the SQP algorithm is given in Equation (35):

On the other hand, we have implemented the scheme of Figure 10 in the Matlab dynamic systems simulator "Simulink," obtaining the following results for state variables and inputs.

$$K = \begin{bmatrix} -33.6207 & -25.2334 \\ 34.6451 & 41.7217 \\ -11.6897 & 49.4783 \\ -19.4097 & -1.0171 \\ -0.2710 & -0.2646 \\ 0.6032 & -6.4790 \end{bmatrix} \tag{35}$$

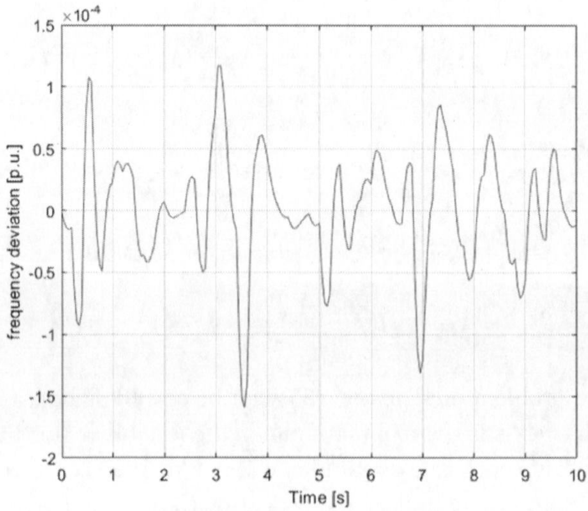

Figure 14. Frequency deviation for the study case power system.

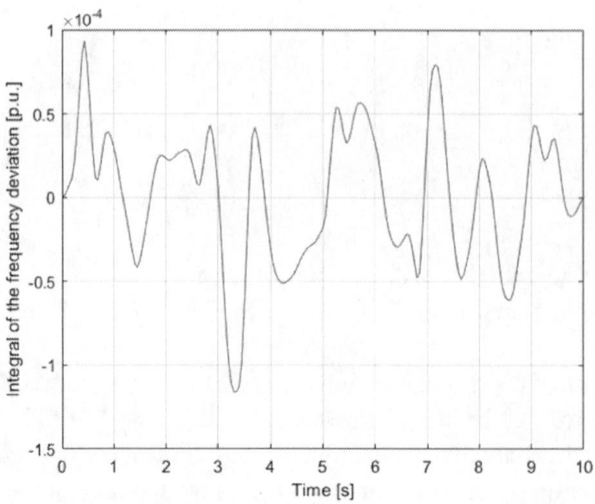

Figure 15. Integral of the frequency deviation for the study case power system.

Figures 14-15, 17-20 illustrate the system states.

From Figure 15, it can be observed that the integral of the frequency deviation varies around [-0.0001 p.u., 0.001p.u.]. Nevertheless, for typical frequency controllers, this value should decrease until cero in a time less than 30 seconds. However, as the generation disturbance (see Figure 13) assumed, in this case, is permanent because it does not disappear in time, this integral value is different from zero. This fact implies that when the renewable wind sources are included in the conventional power systems, the elements on it that are in charge of compensating the load/generation unbalances will be under stress. Moreover, this stress will change according to the penetration degree of the wind sources. In this case, we have considered a conventional generator with a capacity of 128 MVA and wind sources that generate a maximum of 0.8 MVA; i.e., the renewable generation is 0.61% of the conventional generation and the 21.81% of the storage power capacity.

We have also illustrated the absolute frequency error for a power system with a nominal frequency of 60 Hz in Figure 16 with and without considering the anti-windup element.

On it, maximum variations of -0.025 Hz and 0.02 Hz can be appreciated. Furthermore, we can say that this is not a representative frequency deviation since, for example, in Colombia, deviations until ± 0.2 Hz are allowed. However, if a massive inclusion of wind sources is considered, the frequency deviation can become a representative issue.

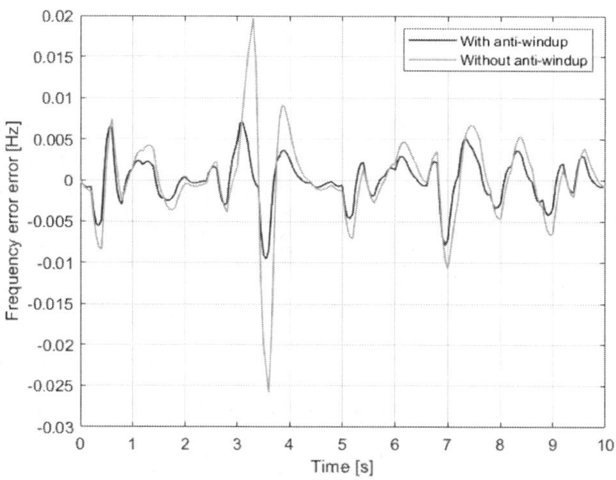

Figure 16. Frequency deviation for control schemes that consider and neglect the anti-windup effect.

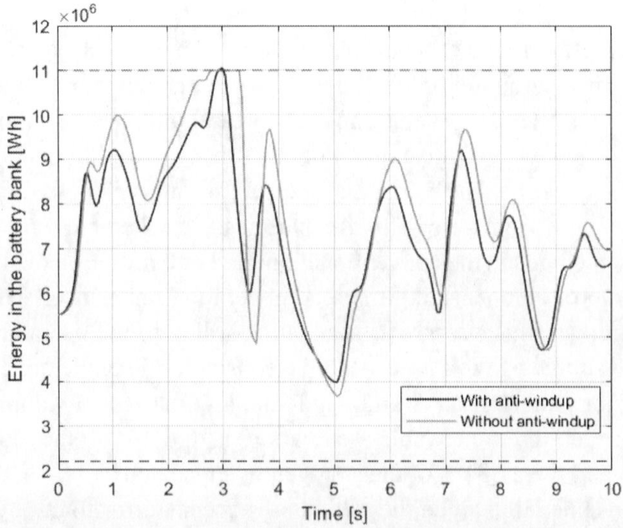

Figure 17. Energy content in batteries for control schemes that consider and neglect the anti-windup effect.

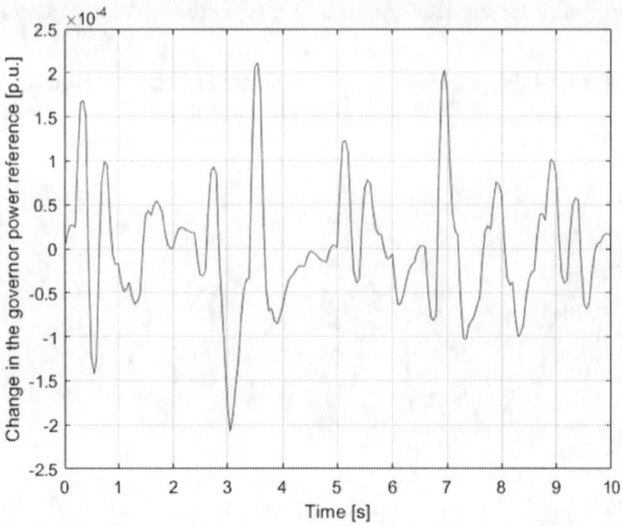

Figure 18. Change in the governor power reference for the study case power system.

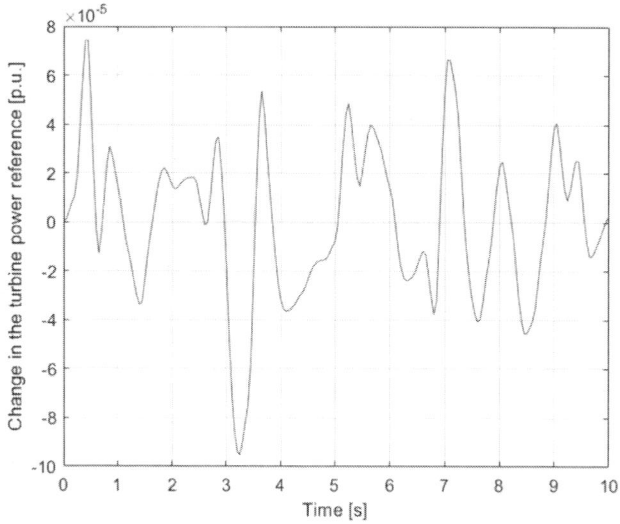

Figure 19. Change in the turbine power reference for the study case power system.

Furthermore, from Figure 16, it can be observed that the anti-windup scheme reduces the frequency deviation since the anti-windup element helps the system compensate for the saturation in the input variables. This can be appreciated in Figure 17 because the batteries reach their limit energy capacity at 3 s in the scheme without anti-windup, which coincides with the time of the maximum frequency deviation.

On the other hand, we have the obtained changes in power reference for the governor, the turbine, and the batteries in Figures 18-19, respectively.

From Figures 18 – 19, it can be observed that the change in the power of the conventional turbine is lower (around 2.5 times) than the required power changes in the governor. This effect occurs because the conventional turbine has a response delay, which is called τ_T, which implies that the turbine will reach the reference change in power given by the governor in a time around $5 \cdot \tau_T$, i.e., 1.5 [s]. In fact, the time response of the conventional turbine is low for the renewable disturbances, and that is the reason why the batteries are scheduled with a higher contribution for the frequency regulation than the conventional generators, as can be appreciated in Figure 20.

On the other side, we also illustrate the integral of the difference between the saturated and unsaturated input actions in Figure 21. On it, it can be observed that at 3 [s], the input reaches its limit value, which means that the energy has touched its maximum value, which is by Figure 17. Moreover, it can be appreciated that the anti-windup element rapidly compensates for this difference.

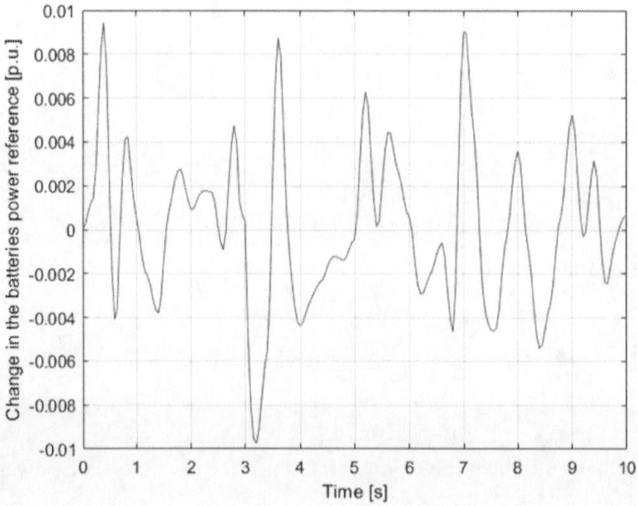

Figure 20. Change in the batteries' power for the study case power system.

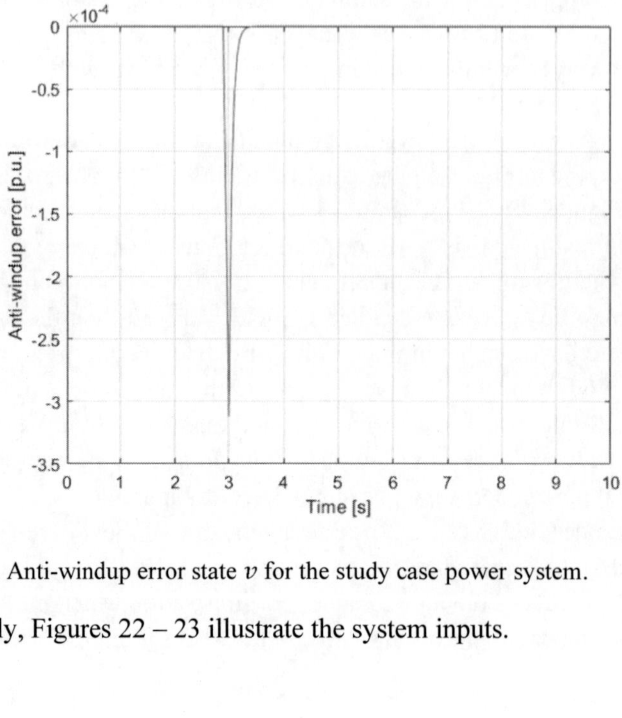

Figure 21. Anti-windup error state v for the study case power system.

Finally, Figures 22 – 23 illustrate the system inputs.

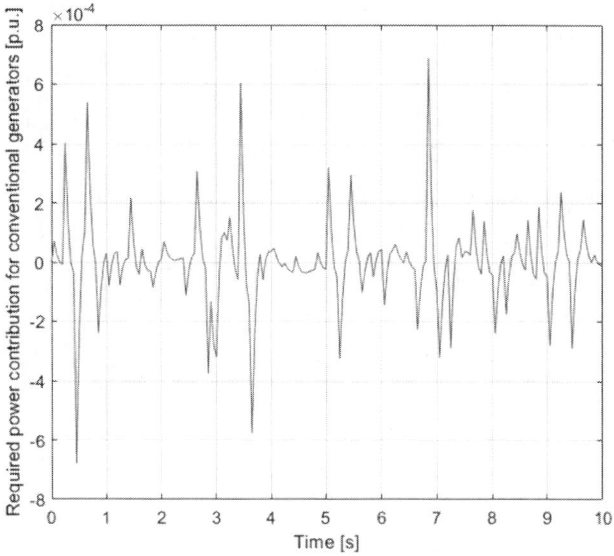

Figure 22. Value of input variable $\Delta P_{ref,g}$ for the study case power system.

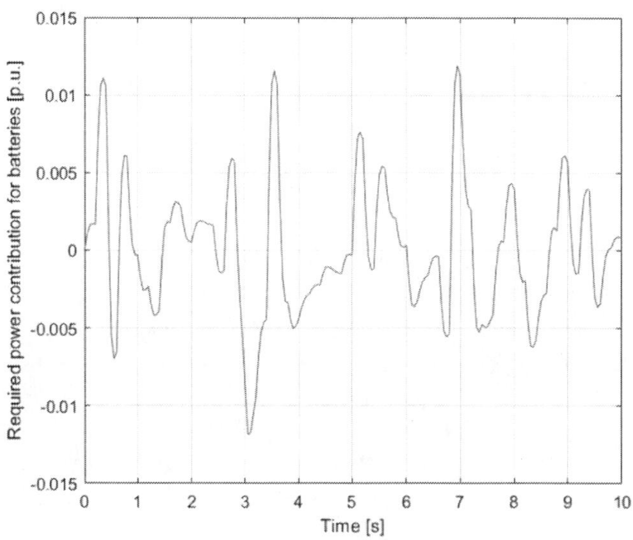

Figure 23. Value of input variable $\Delta P_{ref,b}$ for the study case power system.

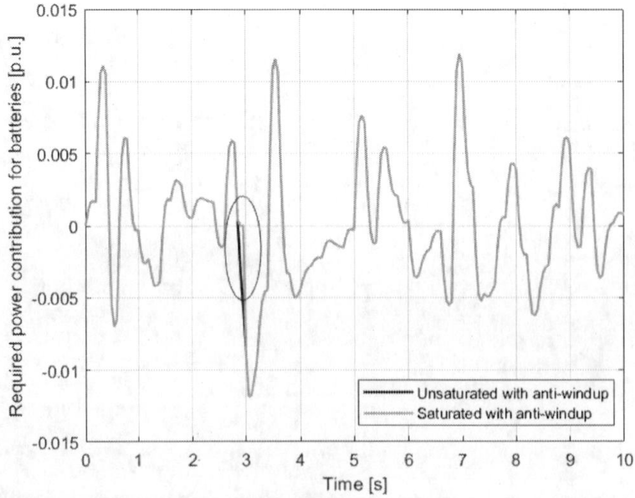

Figure 24. Comparison of the obtained input variable $\Delta P_{ref,b}$ for the study case power system considering the anti-windup.

According to the data depicted in Figures 22 and 23, the batteries assume more than 90% of the frequency deviations because they have a faster response and can assume high-frequency disturbances. However, they are limited by their maximum energy and power capacity. Furthermore, that is the issue we solved with the inclusion of the anti-windup scheme.

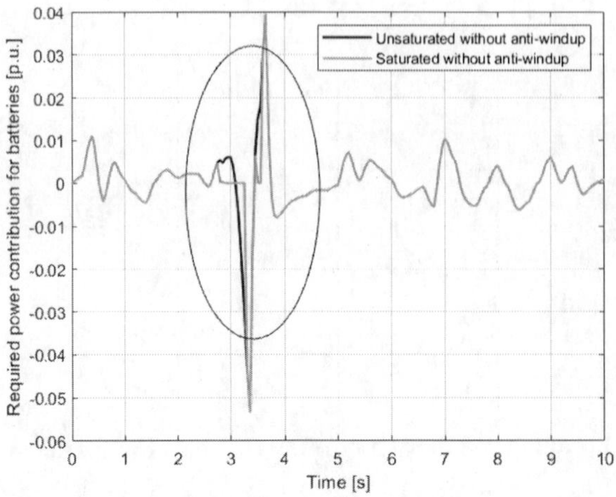

Figure 25. Comparison of the obtained input variable $\Delta P_{ref,b}$ for the study case power system not considering the anti-windup.

In order to clearly show this anti-windup effect, we have also compared the obtained batteries' input after and before the saturation, considering and not considering the anti-windup element.

Figure 24 shows the obtained batteries' input considering the anti-windup element. This figure shows that the difference between the saturated and the unsaturated input is compensated in 0.15 [s].

Although, it does not happen in the case without including the anti-windup, illustrated in Figure 25. In whose case, the compensation of differences between the saturated and unsaturated battery inputs lasts more than 0.85 [s]. Moreover, there is no guarantee that this difference can be attenuated; in many cases, this could lead the system to an unstable operating state.

Conclusion

This chapter presents an optimal feedback control approach to manage the participation of batteries in the frequency regulation loops of conventional systems with the penetration of renewable wind sources. On it, we proposed a nonlinear optimization-based algorithm for tuning the controller parameters in order to obtain the desired frequency response of the closed-loop system based on a pre-defined controller bandwidth and settling time. Furthermore, we deduced this required controller bandwidth and settling time with the power spectrum characterization of the wind generation since it is considered a stochastic generation disturbance for the frequency control system.

Applying this methodology to a proposed study case, we found that, as batteries have a faster response than conventional generators, they are scheduled to assume around 90% of disturbances generated by the wind sources. Under this condition, they can reach their maximum energy or power limits.

Therefore, we find that including these limits in the frequency control scheme is necessary. In this case, we have carried out this task using an anti-windup scheme, which contains a limitation (saturation) for the power supplied by the batteries and for its rate of change, which is calculated based on the estimated state of the charged batteries.

The scheme that we propose is novel concerning the studies reported in the state of the art since it uses control systems with anti-windup, which are widely applied in aerodynamics, and adapt them to the energy management problem in banks of batteries that perform frequency control tasks.

Moreover, we have obtained as a result that the control system with the proposed anti-windup can handle the differences between the saturated and unsaturated battery control input taking it to zero in less than 0.15 seconds. Thus, reducing the maximum deviation of system frequency by 62% compared to a control scheme without anti-windup.

Here we highlight the advantages of the anti-windup technique over other control strategies: low computation burden, it does not require a complex communication scheme to be implemented, and it allows obtaining analytical expressions between the controller parameters and its frequency response.

However, this control technique does not propose a control solution that directly addresses the constraints on the states and inputs of the system. Therefore, in future work, we propose exploring solutions with techniques that directly address the problem constraints, such as dual LQR controllers and MPCs, which can be implemented using the state-space models proposed in this research.

Finally, we highlight another element to be explored in future research: the inclusion of different participation factors for the conventional sources and the batteries in the frequency regulation loop. These factors will allow changing the percentage of the disturbance covered by each element interacting in the frequency control loop, avoiding the excessive participation of the batteries, for example. Moreover, they can be new decision variables in the optimization (31).

Acknowledgments

The author Semaria Ruiz would like to thank Colciencias for its support with the doctoral scholarship of the call number 757 of 2016.

References

Aditya, S. K., and D. Das. 1999. "Application of Battery Energy Storage System to Load Frequency Control of an Isolated Power System." *International Journal of Energy Research* 23 (3): 247–58. https://doi.org/10.1002/(SICI)1099-114X(19990310)23:3 <247::AID-ER480>3.0.CO;2-T.

Arani, Mohammadreza F. M., and Yasser Mohamed. 2017. "Cooperative Control of Wind Power Generator and Electric Vehicles for Microgrid Primary Frequency Regulation."

IEEE Transactions on Smart Grid 3053 (c): 1–1. https://doi.org/10.1109/TSG.2017.2693992.

Baumgarte, Felix, Gunther Glenk, and Alexander Rieger. 2020. "Business Models and Profitability of Energy Storage." *IScience* 23 (10): 101554. https://doi.org/10.1016/j.isci.2020.101554.

Bevrani, Hassan, and Takashi Hiyama. 2011. *Intelligent Automatic Generation Control*. CRC Press.

Biroon, Roghieh A., Pierluigi Pisu, and David Schoenwald. 2021. "*Inter-Area Oscillation Damping via Hybrid LQR State Feedback Control of Large-Scale Battery*," 1–6. https://doi.org/10.1109/naps50074.2021.9449702.

Bossuyt, Juliaan, Michael F. Howland, Charles Meneveau, and Johan Meyers. 2016. "Wind Tunnel Study of the Power Output Spectrum in a Micro Wind Farm." *Journal of Physics: Conference Series* 753 (7). https://doi.org/10.1088/1742-6596/753/7/072002.

Boyd, Stephen, and Craig Barratt. 1991. *Linear Controller Design: Limits of Performance*. Prentice-Hall.

Brooks, AN. 2002. "Vehicle-to-Grid Demonstration Project: Grid Regulation Ancillary Service with a Battery Electric Vehicle." *Regulation*, no. 01: 61. http://www.smartgridnews.com/artman/uploads/1/sgnr_2007_12031.pdf.

Dai, Dan, Tingshu Hu, Andrew R. Teel, and Luca Zaccarian. 2007. *Advanced Strategies in Control Systems with Input and Output Constraints. Lecture Notes in Control and Information Sciences*. Vol. 346. http://www.scopus.com/inward/record.url?eid=2-s2.0-34547146319&partnerID=tZOtx3y1.

Fabozzi, Davide, Nina F. Thornhill, and Bikash C. Pal. 2013. "Frequency Restoration Reserve Control Scheme with Participation of Industrial Loads." *2013 IEEE Grenoble Conference PowerTech, POWERTECH 2013*, 1–6. https://doi.org/10.1109/PTC.2013.6652104.

Fadali, M. Sami, and Antonio Visioli. 2020. *State Feedback Control. Digital Control Engineering*. https://doi.org/10.1016/b978-0-12-814433-6.00009-0.

Greenwood, D. M., K. Y. Lim, C. Patsios, P. F. Lyons, Y. S. Lim, and P. C. Taylor. 2017. "Frequency Response Services Designed for Energy Storage." *Applied Energy* 203: 115–27. https://doi.org/10.1016/j.apenergy.2017.06.046.

Jansson, Samuel. 2019. "Evaluation of KPIs and Battery Usage of Li-Ion BESS for FCR Application." Teknisk- Naturvetenskaplig Fakultet [Faculty of Science and Technology] UTH-Enheten. https://www.diva-portal.org/smash/get/diva2:1356610/FULLTEXT01.pdf

Kinjo, T, T Senju, K Uezato, and H Fujita. 2006. "Output Leveling of Wind Power Generation System by EDLC Energy Storate System." *Electric. Engineer. Jpn.* 154 (4): 1059–66. https://doi.org/10.1002/eej.20181.

Kothare, Mayuresh V., Peter J. Campo, Manfred Morari, and Carl N. Nett. 1994. "A Unified Framework for the Study of Anti-Windup Designs." *Automatica* 30 (12): 1869–83. https://doi.org/10.1016/0005-1098(94)90048-5.

Lalor, G., a. Mullane, and M. O'Malley. 2005. "Frequency Control and Wind Turbine Technologies." *IEEE Transactions on Power Systems* 20 (4): 1905–13. 10.1109/TPWRS.2005.857393.

Ledesma, Pablo. 2008. "*Regulación de Frecuencia y Potencia.*" ["*Frequency and Power Regulation.*"] Universidad Carlos III de Madrid [Charles III University of Madrid].

Leung, Chi Seng, and Peter Meisen. 2005. "*How Electricity Consumption Affects Social and Economic Development by Comparing Low, Medium and High Human Development Countries,*" no. 153: 1–12.

Lu, Chun Feng, Chun Chang Liu, and Chi Jui Wu. 1995. "Effect of Battery Energy Storage System on Load Frequency Control Considering Governor Deadband and Generation Rate Constraint." *IEEE Transactions on Energy Conversion* 10 (3): 555–61. https://doi.org/10.1109/60.464882.

Malik, O. 2015. "Microgrids: Concept and Challenges in Practical Implementation." *IEEE Canadian Review*, 33–35.

Massachusetts Institute of Technology - MIT. 2010. "Feedback Control Systems." *MIT OpenCourseWare.* https://ocw.mit.edu/courses/aeronautics-and-astronautics/16-30-feedback-control-systems-fall-2010/lecture-notes/MIT16_30F10_lec11.pdf.

Momoh, James. 2012. "Smart Grid: Fundamentals of Design and Analysis." Smart Grid: Fundamentals of Design and Analysis, March. https://doi.org/10.1002/9781118156117.

Moore, I., and J. Ekanayake. 2009. "Frequency Response from Wind Turbines." *Universities Power Engineering Conference (UPEC), 2009 Proceedings of the 44th International*, 1–5.

Morren, Johan, Sjoerd W H de Haan, Wil L Kling, and J a. Ferreira. 2006. "Wind Turbines Emulating Inertia and Supporting Primary Frequency Control." *IEEE Transactions on Power Systems* 21 (1): 433–34. https://doi.org/10.1109/ TPWRS.2005.861956.

Nathdatta, Biswa. 2004. "Feedback Stabilization, Eigenvalue Assignment, and Optimal Control." In *Numerical Methods for Linear Control Systems - Design and Analysis*, 343–404.

Pati, Subhranshu Sekhar, and Saroj Kumar Mishra. 2020. "Contribution of Energy Storage Technologies in Load Frequency Control-A Review." *International Journal of Renewable Energy Research* 10 (2): 871–91.

Ramtharan, G., J. B. Ekanayake, and N. Jenkins. 2007. "Frequency Support from Doubly Fed Induction Generator Wind Turbines." *IET Renewable Power Generation* 1 (1): 3–9. https://doi.org/10.1049/iet-rpg:20060019.

Rojas, Osvaldo J., and Graham C. Goodwin. 2002. *A Simple Anti-Windup Strategy for State Constrained Linear Control. IFAC Proceedings Volumes (IFAC-PapersOnline).* Vol. 15. IFAC. https://doi.org/10.3182/20020721-6-es-1901.01089.

Ruiz, Semaria, Julian Patiño, and Jairo Espinosa. 2018. "PI and LQR Controllers for Frequency Regulation Including Wind Generation." *International Journal of Electrical and Computer Engineering (IJECE)* 8 (6). http://www.iaescore.com/journals/index.php/IJECE/article/view/11840.

Saadat, Hadi. 1999. "*Power System Analysis*"

Short, Joe a., David G. Infield, and Leon L. Freris. 2007. "Stabilization of Grid Frequency through Dynamic Demand Control." *IEEE Transactions on Power Systems* 22 (3): 1284–93. https://doi.org/10.1109/TPWRS.2007.901489.

Slootweg, J G, H Polinder, and W L Kling. 2001. "Dynamic Modelling of a Wind Turbine with Doubly Fed Induction Generator." 2001 Power Engineering Society Summer

Meeting Conference Proceedings 1: 644–49. https://doi.org/10.1109/PESS.2001.970114..

Soshinskaya, Mariya, Wina H J Graus, Josep M Guerrero, and Juan C Vasquez. 2014. "Microgrids: Experiences, Barriers and Success Factors." *Renewable and Sustainable Energy Reviews* 40: 659–72. https://doi.org/10.1016/j.rser.2014.07.198.

Swierczynski, Maciej, Daniel Ioan Stroe, Ana Irina Stan, Remus Teodorescu, and Dirk Uwe Sauer. 2014. "Selection and Performance-Degradation Modeling of Limo2/Li 4ti5o12 and Lifepo4/c Battery Cells as Suitable Energy Storage Systems for Grid Integration with Wind Power Plants: An Example for the Primary Frequency Regulation Service." *IEEE Transactions on Sustainable Energy* 5 (1): 90–101. https://doi.org/10.1109/TSTE.2013.2273989.

Thomsen, Sven Creutz. 2006. "Nonlinear Control of a Wind Turbine." Technical University of Denmark.

Ustun, Taha Selim, Cagil Ozansoy, and Aladin Zayegh. 2011. "Recent Developments in Microgrids and Example Cases around the World - A Review." *Renewable and Sustainable Energy Reviews* 15 (8): 4030–41. https://doi.org/10.1016/j.rser.2011.07.033.

Wei, Li, G Joos, and C Abbey. 2006. "Wind Power Impact on System Frequency Deviation and an ESS Based Power Filtering Algorithm Solution." *Proceedings of the 2006 Power Systems Conference and Exposition* 2: 2077–84. https://doi.org/10.1109/PSCE.2006.296265.

Yan, Wenxu, Lina Sheng, Dezhi Xu, Weilin Yang, and Qian Liu. 2018. "H∞ Robust Load Frequency Control for Multi-Area Interconnected Power System with Hybrid Energy Storage System." *Applied Sciences (Switzerland)* 8 (10). https://doi.org/10.3390/app8101748.

Yanagawa, S., T. Kato, Wu Kai, A. Tabata, Y. Yokomizu, T. Okamoto, and Y. Suzuoki. 2001. "Evaluation of LFC Capacity for Output Fluctuation of Photovoltaic Power Generation Systems Based on Multi-Point Observation of Insolation." *2001 Power Engineering Society Summer Meeting. Conference Proceedings* 3: 1652–57. https://doi.org/10.1109/PESS.2001.970323.

Chapter 5

Wind Turbines for Energy Production in Organic Buildings

Daniel Icaza[1,*], David Borge-Diez[2], Semaria Ruiz[3] and Santiago Pulla-Galindo[1]

[1]Catholic University of Cuenca, Ecuador. Center for Research, Innovation, and Technology Transfer CIITT, Lighting Technology Research Laboratory CIITT, GIRVyP Group

[2]Department of Electrical, Systems and Automation Engineering, University of León, León, Spain

[3] Energetic Process Department, Universidad Nacional de Colombia. Medellín, Colombia

Abstract

Organic architecture is a philosophy that ensures harmony between the human habitat and the natural world. Therefore, these buildings are built underground or surface-level. The main objective is to integrate the facilities into the site and take advantage of their innate potential and resources. Energy generation is required on-site to match energy demand; consequently, efficient technologies based on renewable energy generation are the optimal alternative. Small wind turbines are one of the best solutions and can be easily integrated into the site and the building. This chapter analyzes how wind turbines can produce the required energy and assure that organic building principles are achieved. A numerical model for wind systems sizing is presented, showing how this model could be used as an optimization and design tool for such hybrid systems using wind. This solution could be applied to grid-connected organic

*Corresponding Author's Email: dicazaa@ucacue.edu.ec.

In: Wind Turbines
Editors: David Borge-Diez and Daniel Icaza
ISBN: 978-1-68507-974-1
© 2022 Nova Science Publishers, Inc.

buildings, as a self-consumption structure, or to organic architectures in areas far from the public electricity grid.

Keywords: organic buildings, renewable energy, wind turbine, energy conversion

Introduction

Birds' flight, plant shapes, flower colors, and water sprouting, among other wonders of nature, were the inspiration for the multiple and revolutionary inventions that have been made over the years up to now and that human beings keep developing for the benefit of people worldwide [1].

Organic architecture is a philosophy of architecture that promotes harmony between the human habitat and the natural world. Through design, it looks for understanding and integrating the territory, through accurate territorial development plans, into small complementary works in public or private spaces [2]. Built areas and elements such as furniture, housing, and surroundings are propitious spaces to become part of a unified and correlated composition [3]. The most prominent architects under this philosophy are Gustav Stickley, Antoni Gaudí, Frank Lloyd Wright, Imre Makovecz, Javier Senosiain, and Antón Alberts.

However, one of these architects is relevant for this research due to his work with organic constructions. Javier Senosian published his masterpieces in his book [4], inspiring the desire to live in contact with nature. The author showed developments never seen before in this area, indeed wonders in advanced architectural settings. In several projects developed, it is remarkable how its architectural interiors resemble a mother's womb where her baby lives inside. Figure 1 shows the essential outcome of this inspiration.

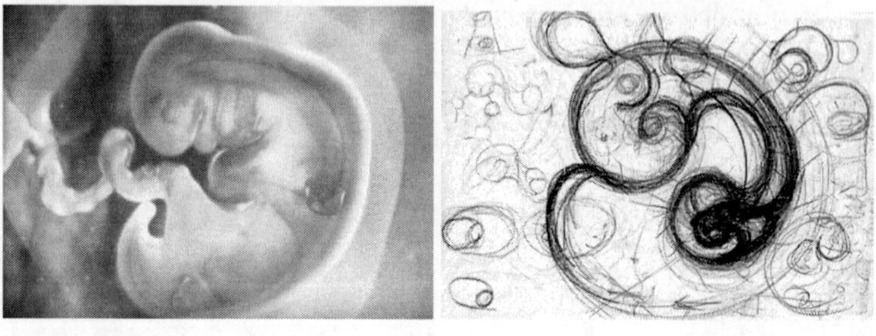

Figure 1. An analogy of a baby in its mother's womb.

The inspiration of maternity is taken to an architectural context where family warmth must prevail, as seen in Figure 1. At the same time, environments are transformed into what is expressed as a mother's womb by Senosian, as shown in Figure 2.

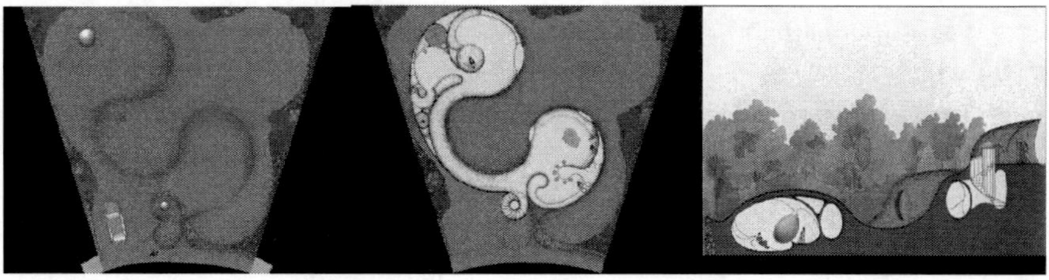

a) (b) (c)

Figure 2. Design of correlated environments. (a) Top view of architecture. (b) Room details. (c) Sectioning.

From a different perspective, each environment is created to appear welcoming and pleasant. The people who inhabit it feel at home with spaces specially designed and inspired by Mother Nature. Subsequently, these spaces are perfectly designed to welcome a family. Aspects related to natural lighting are essential. Basic services for a living must be available, such as drinking water, sewage, electricity, and the Internet.

These spaces are adapted according to the needs of an eco-friendly, physical, and psychological environment, considering men's natural origin. Designs inspire to create particular environments according to local reality. In most cases, it requires as many available spaces as allowed by its location but in harmony with the place.

Some architectural developments are related to nature. They are explained in different research pieces, such as the outcomes presented by Fatma Abdelaal et al. [5]. Besides, research has determined that an adequate indoor microclimate must be maintained, avoiding abrupt variations between indoor and outdoor environments. On the other hand, Ayodele Emmanuel Ikudayisi et al. [6] reviewed the evaluation metrics and the conceptual framework of the integrated design process for green building projects. Marco Ávila et al. [7] go beyond analyzing comfort in interior spaces and focus on studying a tree that provides shade in exterior spaces based on the shadow projected on people who use open spaces. In addition, the available radiation is used and transformed into electrical energy to connect electronic devices. From this viewpoint, it is necessary to focus on the electrical energy supply for these unique buildings. P.J. Antsaklis et al. [8] indicate that the electricity grid is advisable for hybrid energy systems with outside infrastructure.

Another study referring to systems that link natural environments to innovative architectures is the one presented by Xiaoyu Yan et al. [9], with buildings powered only by photovoltaic solar energy with battery storage.

It is crucial to indicate that this type of organic construction has been widely developed from an architectural standpoint. However, from the energy perspective, it is possible to broadly address and establish methodologies that guarantee electrical energy service for buildings. Although there are small developments mentioned above, they are not directly focused on what corresponds to habitable organic constructions.

In this work, the energy supply from renewable sources was reviewed, considering that these constructions must be located in multiple spaces and related to the environment. In real life, access to conventional electrical distribution networks is not always available, and it has been seen that providing a distributed electrical system is necessary. Distributed generation is commonly known as decentralized generation and has several benefits, including low cost, less complexity, and elimination of the interdependencies and inefficiencies associated with transmission and distribution. There are lots of reasons to use distributed generation, such as standby or emergency generation, and the support and great potential of a green energy source using renewable technology. Since these buildings have a relationship with the environment, they will likely be located in open places and outside the city where the electrical service is not necessarily available. Instead, they are remote places disconnected from the network.

Renewable and non-conventional energy generation methods are highly valued for their incredible benefits, and there is a vast variety: wind, solar, hydraulic, biomass, and geothermal. These generation sources are increasingly influencing the electricity market worldwide, considered a replacement for fossil fuels. Different authors emphasize implementing hybrid systems to generate electricity using renewable sources. Among these findings, it is possible to name Miguel Castro Fernandez et al. [10], who developed a model for optimizing a hybrid system in places of residential courts in Cuba. K. Bed et al. [11] presented a genetic algorithm to optimize renewable energy from hybrid wind, and photovoltaic systems. D. Borge-Diez [12] used hydraulic microturbines to solve the lack of energy supply in sectors where open water channels can be used. OM Babatunde et al. [13] analyzed its hybrid system, including a gasoline generator, and determined that costs are reduced. However, the environmental aspect is affected, which leaves it to the discretion of the designers of hybrid systems to accept the option that best suits the reality of each country.

Hybrid 100% renewable energy systems are becoming popular not only as an alternative but as intelligent energy generators in several European countries such as Germany [14], Portugal [15], Spain [16], and France [17]. Alborg University has already been concerned with developing simulation tools such as EnergyPLAN to progressively replace fossil fuels with renewable energies that are becoming more and more affordable from an economic perspective [18]. These systems must complement each other, provide higher quality and more reliable power supply, be independent of the grid, and guarantee electricity in rural areas. In South America, one of the countries that have advanced the most in the last decade is Ecuador. This country changed its Constitution and modified its productive matrix [19]. Among the most important projects carried out, it is possible to name the "Villonaco" project in the Province of Loja [20] and the solar and wind park in the Galapagos Islands [21]. Other projects are underway thanks to legal and financial aspects that have worked in our favor, such as the "Huascachaca" project in the Province of Loja [22] and phase 2 of the "Villonaco" project [23].

The applications that may arise with the intervention of renewable energies can be diverse according to the projections of land use in each locality and current legislation related to energy [24]. In this sense, organic buildings that began to be developed are identified in references [25] and [26]. Additionally, reference [27] presents a design of a system accompanied by communication networks that allows monitoring heritage architectures as a novelty to the existing system. In [28], a photovoltaic solar energy system is considered, but with battery backup to function autonomously. As a result of these studies, it has been considered that organic buildings can be better analyzed considering further developments. However, they require more significant energy guarantees because of their unique design of comfort, which is one of their primary requirements. This study analyzes these wind generation systems in Ecuador. A modern trend in renewable energy systems and their characteristics is introduced to accomplish the abovementioned objective. Under this theoretical framework, the development of renewable energies and electricity systems applied to organic buildings is analyzed. This particular work aimed to structure a small wind machine to be incorporated into hybrid systems and supply these buildings' demands.

Role of Renewable Energies and Benefits

Renewable energies are clean, inexhaustible, and increasingly competitive. They differ from fossil fuels mainly in their diversity, abundance, and potential for usage anywhere on the planet. Above all, they do not produce greenhouse gases (which cause climate change) or polluting emissions [29]. Costs are declining at a sustainable rate, while the general trend in prices for fossil fuels is the opposite, despite their current volatility.

The growth of clean energy is unstoppable, as reflected in the statistics compiled annually by the International Energy Agency (IEA). They accounted for almost half of all newly installed power generation capacity in 2020, when they were the second-largest source of electricity globally, behind coal [30]. According to the IEA, global electricity demand will have increased by 70% by 2040, and its contribution to the consumption of final energy increased from 18% to 24% over the same period, driven mainly by the emerging economies of India, China, Africa, and the Middle East and Southeast Asia [31].

The development of clean energy is vital to deal with climate change and limit its most devastating effects, considering that 2019 was the second warmest year on record [32]. The Earth's temperature has increased by 0.85 °C since the 19th century. Meanwhile, some 1.1 billion people (17% of the world's population) do not have access to electricity [33]. Likewise, 2.7 billion people (38% of the population) use conventional biomass for cooking, heating, and lighting their homes, which poses a severe risk to their health [34]. Renewable energies received significant support from the international community through the Paris Agreement signed at the World Climate Summit held in the French capital in December 2015. The agreement, which entered into force in 2016, establishes a binding global target for the first time. Almost 200 signatory countries committed to reducing their emissions so that the planet's average temperature at the end of this century remains "well below" 2 °C, the limit above which climate change will have more catastrophic effects. The objective is to keep it at 1.5 °C [32].

The transition to an energy system based on renewable technologies will positively impact the world economy and development. According to the International Renewable Energy Agency (IRENA) [35], it will be necessary to double the share of renewable energy in electricity generation to 57% worldwide by 2030 to meet the goals of the Paris Agreement. It requires increasing annual investments in renewable energy from USD 330 billion to

USD 750 billion, thus boosting job creation and growth linked to the green economy [36].

Advantages and Disadvantages of the Use of Renewable Energies

Using renewable energy instead of fossil fuels has several advantages and some of the main benefits of being green.

1. *Renewable energy will not run out*: Renewable energy technologies use resources directly from the environment to generate energy. These sources include the Sun, wind, tides, and biomass, to name just a few of the most popular options. They will not run out. On the other hand, fossil fuel resources are used, and they will become increasingly difficult to obtain; thereby, the cost and environmental impact of extraction will increase [37].
2. *Lower Maintenance Requirements*: In most cases, renewable energy technologies require less overall maintenance than traditional fuel source generators [38]. The generation of technology such as solar panels and wind turbines has few or no moving parts and does not rely on flammable and combustible fuel sources to function. Fewer maintenance requirements translate to more savings in time and money [39].
3. *Renewables save money*: Using renewable energy can help save money in the long run. It will save on maintenance and operating costs [40]. Refuel will not be necessary when technology is generated by the Sun, wind, steam, or natural processes. The money saved by using renewable energy may be based on several factors, including technology. In most cases, transitioning to renewable energy means savings of hundreds to thousands of dollars [41].
4. *Renewable energy has numerous health and environmental benefits*: Renewable energy sources emit little or no greenhouse gases or pollutants [42]. It means a lower carbon footprint and an overall positive impact on the natural environment. During the combustion process, fossil fuels emit large amounts of greenhouse gases, which have been shown to exacerbate rising global temperatures and the frequency of extreme weather events [43]. Using fossil fuels emits greenhouse gases and other harmful pollutants, which lead to respiratory and heart health problems. Renewable energy will

decrease these pollutants' prevalence and create a healthier atmosphere [44].

5. *Renewables are less dependent on foreign energy sources*: Energy is produced locally with renewable energy technologies [23]. The more renewable energy is used, the less imported power will be needed, contributing to energy independence [45].

Renewable energy has many benefits, but it is not all roses regarding renewable energy [13]. These are some of the disadvantages of using renewable energy compared to traditional fuel sources [46]:

a. *Higher Upfront Cost*: While saving money by using renewable energy, the technologies are often more expensive upfront than traditional power generators [47]. There are often financial incentives, such as tax credits and rebates, available to help alleviate the upfront costs of renewable technology [48].
b. *Intermittency*: Although renewable energy resources are available worldwide, many of these resources are not available 24/7 throughout the year. Some days can be windier than others, the Sun does not shine at night, and droughts can occur eventually [49]. There may be unpredictable weather events that disrupt these technologies [50]. Fossil fuels are not intermittent and can be turned on or off anytime.
c. *Storage capacities*: Due to the intermittency of some renewable energy sources, there is a great need for energy storage [51]. While storage technologies are available today, they can be expensive, especially for large-scale renewable power plants. It is worth noting that energy storage capacity increases as technology advances, and batteries become more affordable as time goes on [52].

When it comes to renewable energy, the positives outweigh the negatives. Transitioning to renewable energy at a personal, corporate, or government level will save money and promote a cleaner, healthier environment for the future [53].

Non-Renewable Energy

Non-renewable energy is derived from finite resources that are not replaced quickly enough to keep up with the consumption rate. In perspective, non-renewable energy sources will not be replenished for many human lives [54]. Most non-renewable energy sources are fossil fuels such as crude oil, coal, and natural gas. However, nuclear power, used primarily to produce electricity, is generally non-renewable [55]. Fossil fuels take their name from their origins, formed from ancient plants and animals buried, carbon-rich remains. Most fossil fuel deposits were formed between 540 and 65 million years ago [56]. Over time, the underground compression of these organisms mentioned above gradually gave rise to a mineral called kerogen, which then combined with geothermal heat to form fossil fuels [57].

Fossil fuels are partly harmful because they release carbon dioxide into the atmosphere when they heat up, ultimately upsetting the carbon balance on land, ocean, and air. Due to the formation process of fossil fuels, the carbon found in them has been sequestered underground for millions of years; when this carbon is released into the atmosphere, temperatures rise too quickly for organisms to keep up and adapt. Hence the term "global warming" first entered the public lexicon in a 1975 journal article by climate change researcher, scientist, and pioneer Wallace Smith Broecker [58].

Many decades before Broecker issued the first global warming alarm, the unsustainable practice of burning fossil fuels, which releases carbon dioxide, generated most of the energy needed to "power" economies and the highly industrialized lifestyles of modern society. Fossil fuels have been essential in areas ranging from public and private transportation to domestic electricity and plastics production [53]. The planet has more than paid for our globalized economy as there is an ever-increasing focus on efficiency and production. These results could be evidenced in numerous areas: air and water pollution, climate change, and the health and environmental effects associated with practices such as fracking, a controversial method of extracting oil and natural gas [59].

Although the transition from fossil fuels and other non-renewable energy to clean energy is ongoing, most of the power in human society still comes from non-renewable resources. Fossil fuels dominate globally, providing about 80 percent of the total energy used yearly. Some researchers estimate closer to 85 percent [60]. For illustration, in the high-consuming United States, primary energy sources are 89 percent non-renewable. There are three main types of fossil fuels: coal, oil, and natural gas [61].

Coal is a black or brown combustible rock mainly used as fuel with subtypes such as "peat" or "anthracite." It was also the main driver of the industrial revolution of the 18th and 19th centuries, when the foundations were laid for many modern manufacturing processes. Even though it is abundant, reliable, and inexpensive, coal is highly polluting when burned [62]. Coal mining is also dangerous, as miners face constant exposure to the toxic dust and the imminent threat of on-site explosions.

Also known as crude oil, it keeps much of the world's wheels turning. About half of its supply of this liquid fossil fuel is converted to gasoline [15]. *Oil* is an easily transportable (non-renewable) energy source that allows people to keep moving. Nevertheless, the downsides of oil today cannot be ignored: Burning gasoline releases toxins into the air humans breathe [42]. It is extracted through various drilling methods that carry many inherent risks, including potential oil spills, which can devastate the environment. Natural gas, made mostly of methane, is found underground in deposits. The main uses of natural gas are heating, cooling, and electricity generation [52]. Natural gas is a "cleaner" fossil fuel than oil and coal since it releases fewer pollutants into the air when burned (carbon dioxide and water vapor only). However, natural gas extraction processes can lead to many environmental problems and disasters, such as the increased possibility of small earthquakes and water contamination [63].

Nuclear power is obtained from radioactive elements. Although it is not a fossil fuel, nuclear power is generally classified as a non-renewable form of energy, mainly because most nuclear reactors rely on uranium as a fuel source (There is a limited supply of uranium in the world) [64]. Used to produce electricity, nuclear power has some critical advantages over fossil fuels. It contributes significantly less to air pollution and carbon emissions. An American Chemical Society study even indicated that between 1971 and 2009, nuclear power helped prevent 1.84 million premature deaths related to air pollution and 64 billion tons of greenhouse gas emissions [6]. However, nuclear power remains highly controversial due to the radioactive threats that nuclear power plants pose to people and the environment, as seen in devastating nuclear accidents such as the 1986 Chornobyl disaster, which took place on the outskirts of the Pripyat town in the Ukrainian SSR.

Despite their limited supply and evident devastation potential, the world remains dependent on non-renewable energy, particularly fossil fuels. Despite all their flaws and dangers, these are energy-dense (instantaneous) and relatively inexpensive to process. In addition, the storage methods are simple and can be shipped anywhere in the world. While many governments, corporations, and other vital players now recognize the urgency of switching to alternative energy sources and reducing CO_2 emissions, doing so effectively will require a long-term, sustained, and costly overhaul of existing systems [60]. International solidarity and collaboration are crucial to fight against climate change and ensure a clean energy future in this transition. In Ecuador, non-renewable energy sources derived from natural gas and oil in 2020 registered an energy production of 3,413.21 MW of nominal power throughout the territory [58].

Currently, in the country, the Non-Renewable energy generation plants are distributed in four regions: five provinces in the Amazon and Insular, six provinces in the Sierra, and six provinces on the Coast. By 2020, the Amazon and coastal areas will have the highest capacity with 1,720.09 MW and 1,409.96 MW, respectively [30, 65]. However, the province of Guayas has various thermal power plants with a total power of 805.21 MW. Similarly, in Orellana, a thermal power plant with a nominal power of 783.18 MW corresponds to self-generating companies in the oil sector [59]. Finally, by 2020, the MCI thermal plants (internal combustion engines) had a nominal power of 2,029.74 MW, while the turbo gas and turbo-steam plants had a nominal power of 921.85 MW and 461.63 MW, respectively [66].

Renewable Energies in Ecuador

Article 413 of the Constitution of the Republic of Ecuador indicates [19] that: The State promotes energy efficiency, development, and use of environmentally friendly practices and technologies clean and healthy, as well as renewable, diversified, low-impact energy, and does not jeopardize food sovereignty, the ecological balance of ecosystems or the right to water [19]. Backed on this chapter of the book, the Ecuadorian government, through its entities, has begun to expand and encourage the use of clean and efficient energies, as well as the responsible exploration and exploitation of hydrocarbons, opting for a policy aimed at the efficient use of renewable resources and thus fortify its energy matrix [67].

Planning a new matrix of Ecuadorian energy is based primarily on generating a balance through government policies, construction of projects for energy generation plans stipulated in the short and medium-term, the institutionalization of technical capacities, and all the factors involved, such as energy policy, subsidy law, and public and private institutions. This matrix has varied over time, always maintaining three predominant factors in energy generation, oil, biomass, and hydroelectricity. The most significant changes started in 1970 when there was the modernization of the country due to the exploitation of the oil industry and the growth in fossil fuel consumption, replacing the burning of firewood that was the main energy-generating factor up to that era [68]. This modernization opened a critical gateway to international capitalism due to the significant income generated in Ecuador from oil exploitation, becoming an essential attractiveness for foreign bank investment [69]. Also, at this time, the construction of a new energy source began; one of the most essential and significant hydroelectric plants that Ecuador has, the Paute Hydroelectric Plant, which ended in 1991, contributed a large amount of energy hydroelectric to the Ecuadorian energy system [70].

Between 1980 and 1990, the energy consumption rate did not increase much compared to other decades [68]. It coincides with the country's financial crisis at the beginning of this decade due to oil prices collapsing, international indebtedness, and mismanagement of money [41]. Other energy sources were added in the following decades, contributing to the energy system. Projects were built on photovoltaic and wind power in 2005. They were exciting solutions for electricity supply in many locations, reducing production costs.

In 2011, the energy demand grew faster (more than 6.10% per year) than the economic growth rate (4.37%) compared to 2007. The government implemented a project to regulate energy consumption granting a subsidy to residential users with consumption of up to 110 kWh Sierra region and up to 130 kWh in the coastal, eastern and insular regions per month. This project was shaped by the "Dignity Tariff" embodied in Executive Decree No. 451-A (2007). Until this year, there were 207 generation plants, of which 88 were incorporated into the National Interconnected System [71].

As a campaign promise of the former President Rafael Correa, the new Constitution was made and was in force until 2008. During this time, it was proposed among its objectives to make changes in the country's energy matrix, promoting nature rights; Ecuadorian and Bolivian Constitutions are the only ones that give these rights [72]. Implementing measures that help solve climate change mitigation reduces greenhouse gas emissions into the atmosphere, deforestation of forests and jungles, and protects the affected population, taking measures and precautions for their safety and conservation.

Likewise, the National Plan for Good Living was elaborated [73], expressing the intention of transforming specific strategic sectors into environmental and technological ones. Between 2012 and 2014, many projects were built for hydraulic power generation [67]. Among them are La Merced de Jondachi, Baeza, Angamarca, Turtle, Crespi, Chontal, Chirapi, Río Zamora, Verdigris with a total investment of $5,365,270.00.

The former President Lenin Moreno replaced Rafael Correa, and his administration brought a setback in the area. Moreno did not opt for re-election. Guillermo Lasso's administration, which came to power in 2021 had good intentions to create an ecological transition with a view to 2050 and leave fossil fuels behind. There are still no clearly defined and financed projects in this government.

Implementation of Energy Efficiency Policies

Energy efficiency is the fastest and most cost-effective tool for decoupling economic growth since the increase in energy consumption. It also reduces greenhouse gas emissions by lowering the energy needed to achieve a particular amount of energy service [71]. However, efficiency policies consist of technological changes: considering the parametric structure of transport consumption, the implementation of policies is based on changes in technical parameters according to the different categories; these parameter changes are the same for all energy efficiency scenarios.

In general, Ecuadorian society must initiate a process of economic transition based on available resources, considering that its oil horizon is unfavorable and aggravated by the growing energy demand [74]. This transition must start with policies focused on using available hydroelectric resources, which until now have contributed to reducing dependence on fossil fuels in electricity generation. According to the model, there will be a 10-year electricity generation surplus; the maximum surplus will be 60% over five years, reduced to 19% by 2030. The leading cause of this reduction is the increase in household electricity consumption due to the introduction of household appliances. Between 2010 and 2015, the average electricity consumption of households was 7,000 GWh, which would increase to 16,000 GWh between 2025 and 2030.

Therefore, the results show the importance of defining economic and energy policies to use this available resource. It mainly focuses on sectors that generate added value, focused on the industrialization process, and not only allocating it to household consumption, which does not cause added value in the domestic economy.

However, the current electricity surplus management policy focuses on regional exports. Based on the extension plans of neighboring countries, this policy would be affected in the short term, leading to the underutilization of installed capacity. Furthermore, governments have implemented little regional energy integration; there are different opportunities, given the existing complementarity between countries and the existence of common problems and similar markets. Beyond traditional measures to reduce energy consumption, save costs, and lower atmospheric emissions, policies increasing energy efficiency also have multiple economic and social benefits.

It contributes significantly to energy security by reducing dependency on imported energy from an energy supply perspective. In addition, the decrease in energy consumption through greater efficiency reduces the inflationary pressures caused by the increase in international prices of energy raw materials. Finally, industry-wide productivity gains from increasing reliability of operations and processes can lead to positive macroeconomic impacts, from boosting economic activity to generating more employment or having a positive impact on health and well-being. Implementing energy efficiency policies is currently practiced in most countries, and the number of such approaches has grown substantially since 1990 [75]. The focus of these policies has also changed, and they are now more geared toward energy efficiency than concentrating solely on energy conservation. In addition, they are highly influenced by the growing importance given to climate change goals. Energy-efficient technologies can help reduce environmental damage associated with energy use.

Although, as noted above, companies may have incentives to invest in energy-saving technologies to reduce costs, investment in energy efficiency may be below the social optimum without public policy support due to market failures [76]. In particular, the lack of information on available technologies and the uncertainty of energy investments justify the use of public policies.

Some policies are created to improve energy efficiency using various measures and instruments to encourage companies and industries to reduce greenhouse gas emissions and invest in energy-saving technologies. In environmental economics, a common distinction is made between command and control (regulations that set limits and standards) and market-based instruments (subsidies, environmental taxes, tradable emission permits, or tax credits) [77].

After analyzing the aspects of renewable energies and their role worldwide and in Ecuador, a study that involves developing low-power wind turbines is presented, in this case, 800W. These turbines will allow the formation of hybrid systems for isolated places in Ecuador and are fully coupled to supply organic infrastructures, which is the object of analysis of this study.

Simplified Mathematical Model of the Wind Turbine

Figure 3 shows the simplified diagram of the schematized wind turbine. A vane is included to identify the direction of the most significant wind flow and self-position and generate the most effective potential energy [78]. This chapter presents the process of dimensioning the main elements of the wind turbine on all the elements subjected to a rotation to assemble all the elements, including the wind turbine DIAWIND-A2. It was built to supply power to rural homes and can self-position after the census of the direction of greater wind flow through a weather vane. Wind turbines such as this one can be produced in the series for remote areas of Ecuador and neighboring countries. This research can be improved and make values as eligible as possible [79]

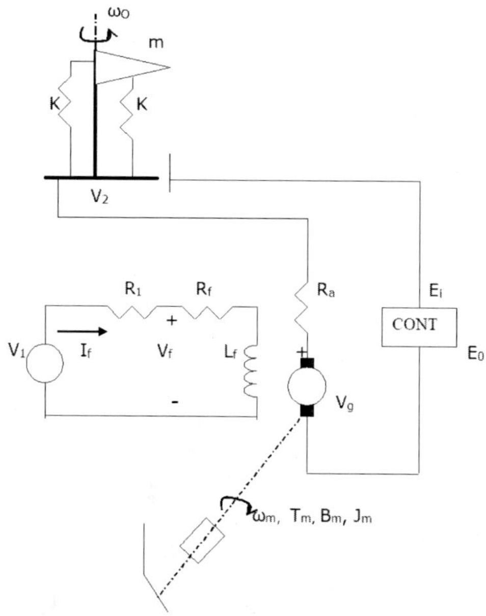

Figure 3. Simplified mathematical model of a wind turbine.

Where:

ω_O	= Angular velocity produced by the wind to wind vane.
k	= Spring constant given in {N/m}.
V_2	= Contact voltage drop.
R_g	= Generator resistance (Alternator).
V_g	= Generated electromotive force.

CONT = Controller selected.
E_i = Input voltage to the controller.
E_0 = Output voltage of the controller.
ω_g = Angular velocity produced in the generator.
ω_1 = Angular speed of entrance to the blades.
R_f, L_f = Stator-related impedance.
V_1 = Utility voltage max. 12V.
R_1 = Stator input resistance.
N_1, N_3 = number of turns of gears driven.
N_2, N_4 = number of turns of conductive sprockets
K_s = Sensor constant (wind vane).

Once the mathematical model is determined, it becomes easier to analyze the system and establish the most representative parameters to meet the generation conditions. Next, the equations that govern the time domain are presented, and they are passed to the Laplace domain and obtain the block diagram, Eq. (1) to Eq. (17)

$$\omega_0 = \sqrt{\frac{k}{m}} \tag{1}$$

$$\omega_0 = \sqrt{\frac{10N/m}{0.4kg}} = 5rad/s$$

$$V_2(t) = K_b\omega_2(t) \tag{2}$$

$$V_2(s) = K_b\omega_2(s) \tag{3}$$

$$V_f(t) = R_f i_f + L\frac{di_f}{dt} \tag{4}$$

$$V_f(s) = (R_f + sL_f)I_f(s) \tag{5}$$

$$V_g(t) = K_g i_f(t) \tag{6}$$

$$V_g(s) = K_g I_f(s) \tag{7}$$

$$V_g(t) = R_g i_a + V_2(t) + K_c E_0(t) \tag{8}$$

$$V_g(s) - V_2(s) = R_g I_a(s) + K_c E_0(s) \quad (9)$$

$$V_1(t) = R_1 i_f(t) \quad (10)$$

$$V_1(s) = R_1 i_f(s) \quad (11)$$

$$T_m(t) = T_L(t) + T_e^1(t) \quad (12)$$

$$T_e^1(t) = -\frac{N_2}{N_1}\frac{N_4}{N_3} T_e(s) \quad (13)$$

$$T_L(t) = J_{eq}\frac{d\omega_m}{dt} + B_{eq}\omega_m \quad (14)$$

$$T_L(s) = (B_{eq} + J_{eq}s)\omega_m(s) \quad (15)$$

$$B_{eq} = B_m + \left(\frac{N_2}{N_1}\frac{N_4}{N_3}\right)^2 B_L \quad (16)$$

$$J_{eq} = J_m + \left(\frac{N_2}{N_1}\frac{N_4}{N_3}\right)^2 J_L \quad (17)$$

Blocks Diagram

Figure 4 shows the block diagram obtained after analyzing each part of the mathematical model.

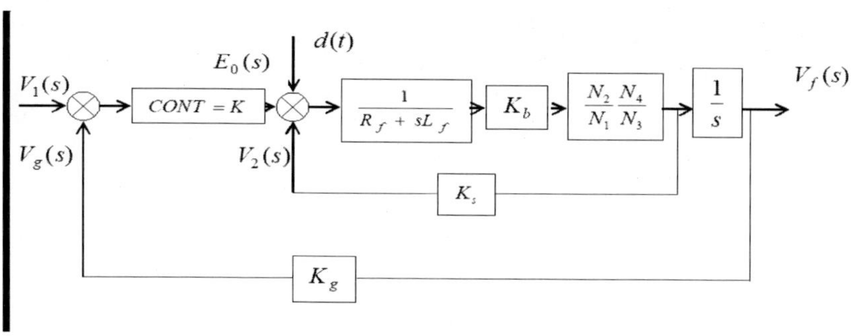

Figure 4. Block diagram.

Transfer Function

Using the principles of block reduction to feedback systems related to an output C(s) and input R(s), the following mathematical relation Eq. (18) is used [80]:

$$\frac{C(s)}{R(s)} = \frac{G(s)}{1+G(S)H(s)} \tag{18}$$

By applying Eq. (18) to our system [81], first between $V_3(s)$ y $V_2(s)$, Eq. (19) is obtained:

$$\frac{V_3(s)}{E_o(s)} = \frac{\frac{1}{R_f+sL_f}\frac{1}{Beq+Jeqs}K_b\frac{N_2N_4}{N_1N_3}}{1+K_s*\frac{1}{(R_f+sL_f)}\frac{1}{(Beq+Jeqs)}K_b\frac{N_2N_4}{N_1N_3}} \tag{19}$$

By organizing terms, Eq. (20) is obtained:

$$\frac{V_3(s)}{E_o(s)} = \frac{K_bN_2N_4}{N_1N_3(R_f+L_fs)(B_{eq}+J_{eq}s)+K_bN_2N_4K_S} \tag{20}$$

Finally, after applying Eq. (18) between the output $V_f(s)$ and the input $V_1(s)$, the transfer function is determined with Eq. (21):

$$\frac{V_f(s)}{V_1(s)} = \frac{K_bN_2N_4[CONT]}{[N_1N_3(R_f+L_fs)(B_{eq}+J_{eq}s)+K_bN_2N_4K_s]s+[CONT]K_bN_2N_4K_g} \tag{21}$$

Subsequently, after evaluation, Eq. (22) is elicited:

$$i = \frac{N_2N_4}{N_1N_3} = \frac{10.10}{20.300} = 1/60$$

$L_f = 0.2H$
$R_f = 4\Omega$
$K_s = K_g = 1$
$K_b = 1$ V/rad/s
$B_{eq} = 0.6$ N-m/rad/s
$J_{eq} = 1.2$ kg-m^2

$$\frac{V_f(s)}{V_1(s)} = \frac{[CONT]}{[60(4+0.2s)(0.6+1.2s)+1]s+[CONT]} \tag{22}$$

Organizing terms are determined in Eq. (23):

$$\frac{V_f(s)}{V_1(s)} = \frac{[CONT]}{14.4s^3+295.2s^2+145s+[CONT]} \quad (23)$$

When evaluating Eq. (20), is obtained Eq. (24):

$$\frac{V_3(s)}{E_o(s)} = \frac{1}{14.4s^2+295.2s+145} \quad (24)$$

Then in Figure 5, it is possible to evaluate the block diagram to the maximum, remaining to determine the CONT controller.

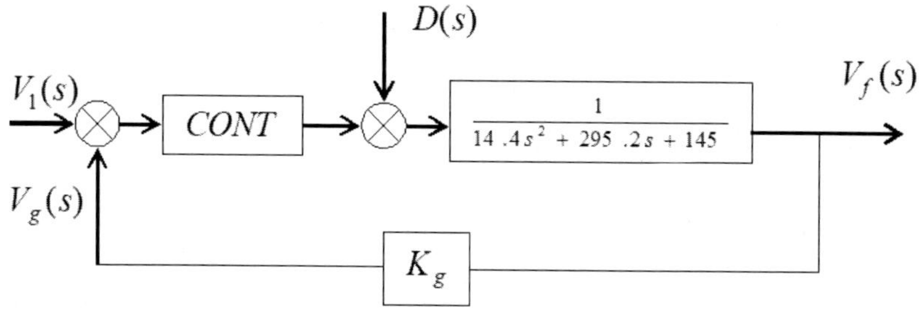

Figure 5. Simplified block diagram.

Starting from (23), (25) is determined:

$$V_1(s) = \frac{12}{s}$$

$$V_f(s) = \frac{[CONT]}{14.4s^3+295.2s^2+145s+[CONT]}\frac{12}{s} \quad (25)$$

D(s) = Possible disturbance generated by turning the wind vane.

$$\frac{V_f(s)}{D(s)} = -\frac{1}{14.4s^3 + 295.2s^2 + 145s}$$

Taking the transfer function shown in Eq. (23), the appropriate controller must be determined. So, it is necessary to perform the following analysis:

It is tested with different controllers, and the one that manages to satisfy is a $[PI]$ for $[CONT] = K_p + \frac{K_I}{s}$:

$$\frac{V_f(s)}{V_1(s)} = \frac{K_p + \frac{K_I}{s}}{5s^3 + 202.5s^2 + 101s + (K_p + \frac{K_I}{s})}$$

$$V_f(t=\infty) = \lim_{s\to 0} s\, V_f(s) = \lim_{s\to 0} \frac{K_p s + K_I}{5s^4 + 202.5s^3 + 101s^2 + K_p s + K_I} = \frac{K_I}{K_I} = 1$$

The characteristic equation is:

$$5s^4 + 202.5s^3 + 101s^2 + K_p s + K_I = 0$$

It satisfies the system's stability since all the equation coefficients are different from zero. Therefore, [PI] was selected, which is determined in Eq. (26):

$$\frac{V_f(s)}{V_1(s)} = \frac{K_p s + K_I}{14.4s^4 + 295.2s^3 + 145s^2 + K_p s + K_I} \quad (26)$$

With values of $K_p = 1$ and $K_i = 0.5$, the transfer function is finally evaluated in Eq. (27):

$$\frac{V_f(s)}{V_1(s)} = \frac{s + 0.5}{14.4s^4 + 295.2s^3 + 145s^2 + s + 0.5} \quad (27)$$

As shown in Figure 6, the poles are in the left plane, concerning the imaginary axis. So, the system is stable.

Next, Figure 7 shows the prototype designed to be coupled to other renewable energy sources and supply energy to organic buildings to form hybrid systems that include wind energy through these small-sized machines [79].

The main elements that make up the wind turbine must be mechanically designed, considering the wind turbine's efforts. It is essential to take care of every detail. Above all, comparisons must be made between calculations, design, and construction on several occasions since these three elements must always go together. The construction processes employ machining by chip removal using flat or cylindrical machine tools, such as the parallel lathe, milling, filling, or grinding machine. These processes obtain parts with a previously designed geometric configuration. Proposing to make a wind turbine with a vision of mass production to supply organic buildings requires a previous analysis through prototypes before making investments to produce a model available to society in the future. Below, in Figure 8, are some details of the assembly.

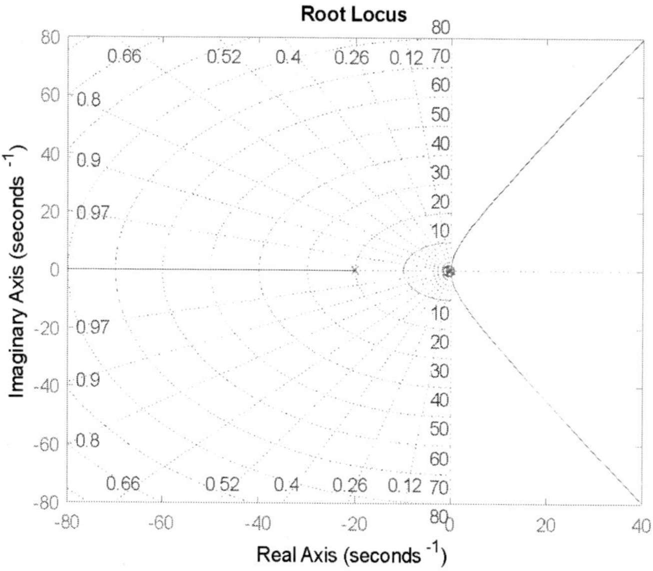

Figure 6. Simulation of the LGR of the wind turbine.

The entire structure of the wind generator will be elevated on the tower. The previously dimensioned vertical axis is housed in the center, and the system will be oriented according to the maximum wind speed. The entire frame rests on a bearing with its respective safety devices and is in a position to rotate around its axis.

Once the wind turbine is positioned according to the maximum wind current, the energy is generated, passes through a charge controller, and is stored in 12V batteries in direct current [52].

After its construction, the operational tests were carried out in the town of Quingeo in the Canton of Cuenca. Additionally, the results of the calculations were made, and the measurements were compared [82].

Figure 7. DIAWIND-A2 wind turbine prototype.

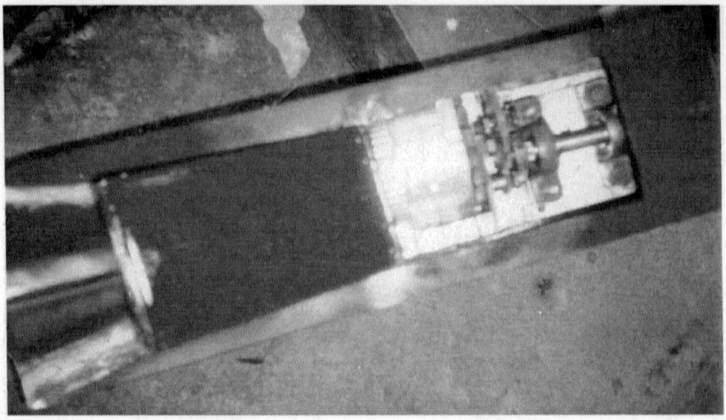

Figure 8. Wind turbine assembly.

The power of each turbine is given by Eq. (28):

$$P_{WT} = \frac{1}{2} \cdot C_p \cdot \rho_{air} A \cdot v^3 \cdot \eta_{WT} \tag{28}$$

Where P_{WT} is the power produced by the sweep of the wind turbine blades, C_p is the Beltz coefficient, $\rho_{air}$is the density of the air, A is the area swept by the wind turbine blades, v is the nominal speed of the wind turbine, and η_{WT} is the efficiency of the wind turbine.

Considering the internal efficiency of the wind turbine, the Eq. (29) can be written as follows:

$$\eta_{C1} = \eta_{fmec} \cdot \eta_g \cdot \eta_{mp} \quad (29)$$

Where η_{fmec} and η_g are the friction and generation efficiencies, while η_{mp} is the efficiency in the speed multiplication box.

Eq. (28) is taken into consideration and expressed in single-phase power in Eq. (30) to determine the output power:

$$P_{1f} = \sqrt{3}.\eta_{c1}.U_{line}.I_{line}.cos\varphi \quad (30)$$

Where P_{1f} is the single-phase power, $_{Elaine}$ is the line current, $cos\varphi$ represents the power factor, and η_{c1} is the conversion efficiency of the wind system.

As one can identify, they are two technically realized wind turbine designs. Still, they are also two varieties that will allow the potential client to decide on one or the other due to its shape. This way, various types can be generated according to color and shape.

Economic Analysis

The summary of income and expenses expected in the first four years according to the scenario is proposed in Table 1.

Figure 9 details the expected income in the first four years. It is crucial to indicate that wind machines are designed to be implemented in organic buildings and homes to expand the target market.

Table 1. Projected income over time

Year	P ($)	Q	I ($)
1	1100	900	990000
2	1100	970	1067000
3	1000	1150	1150000
4	1000	1240	1240000

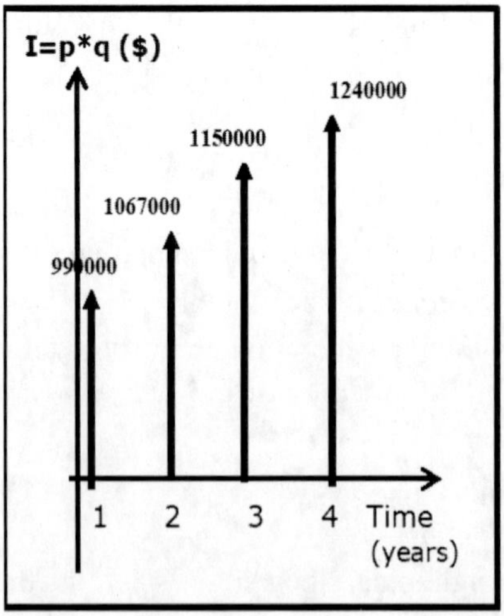

Figure 9. Expected income in the first four years.

Figure 10 details the annual investments to be made. Year 1 requires the most resources due to the necessary initial investment.

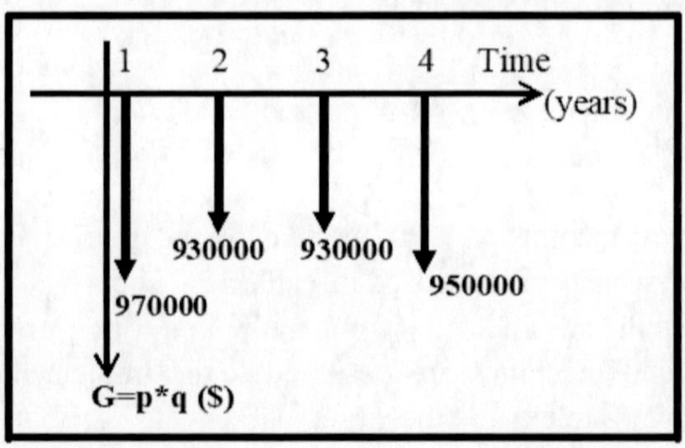

Figure 10. Investment flow in the first four years.

Figure 11 shows the annualized profitability according to the scenarios proposed in the study.

After construction, operating tests were carried out in the town of Quingeo in the Canton of Cuenca - Ecuador, as shown in Figure 12. Comparisons with the proposed model are also made. The curves are represented as follows:

$A = 7.068m^2$

$C_P = 0.5$

$\rho_{air} = 1.01\ Kg.m^{-3}$
$\eta_{aer} = 0.5$

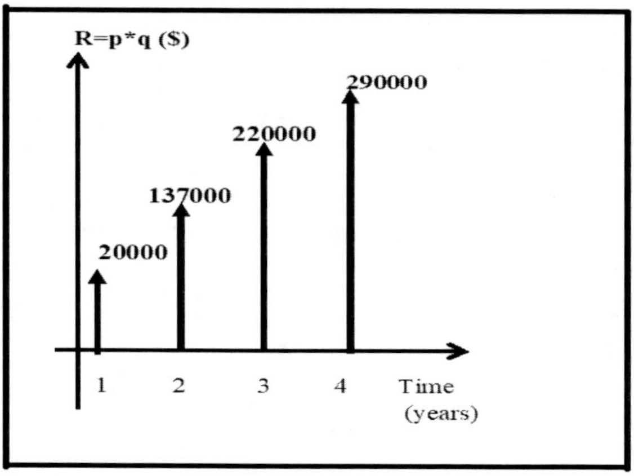

Figure 11. Profitability analysis projected four years after its implementation.

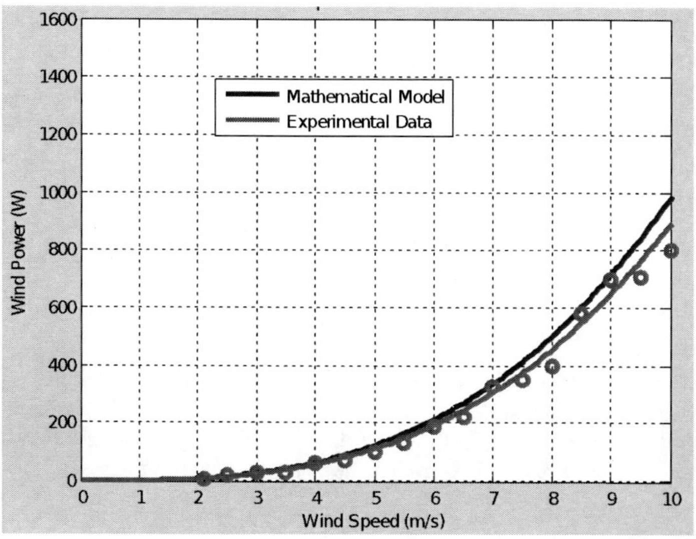

Figure 12. Comparative curves between the theoretical mathematical model and the experimental curve of the DIAWIND-A2 wind turbine.

Conclusion

The production of wind turbines in series is possible in our environment. It significantly impacts the population since it would become accessible, especially to the rural sector. However, there are some fears to overcome when accepting such a technology; breaking these barriers may take time. Once this aspect is resolved, this technology would cover many energy needs in Ecuador and the region.

The machine tools for constructing these wind turbines are accessible; the workforce can also be arranged. Investment is key, as well as promotion to exploit it massively. Also, these wind turbines can capture the maximum wind currents and generate the maximum possible electrical energy.

As we can see, despite being a relatively small machine, it is possible to start generating energy with relatively small wind speeds of around 2.5 m/s. This way, a reasonably substantial range of these machines in various rural localities will be beneficial for installation in the vicinity of the organic buildings proposed in this document. Over time, modifications can be made to be massively implemented at the residential level and form hybrid systems recommended in the literature review under a territorial order that should be legislated.

Wind energy generation is clean energy, abundant in Ecuador, and easily coupled to other sources. The power generated can be stored in a battery bank to be supplied during non-generation hours.

References

[1] Li, Y., Y. Rong, U.M. Ahmad, X. Wang, J. Zuo, G. Mao, A comprehensive review on green buildings research: bibliometric analysis during 1998–2018, *Environ. Sci. Pollut. Res. Int.* (2021) 1–19. https://doi.org/10.1007/s11356-021-12739-7.

[2] Kirsch, G., Developing A Sustainable Land Ethic In21st Century Cities, in: *ARC* 2012, 2012. https://doi.org/10.2495/ARC120131.

[3] Kadaei, S., S.M. Shayesteh Sadeghian, M. Majidi, Q. Asaee, H.H. Mehr, Hotel Construction Management considering Sustainability Architecture and Environmental Issues, *Shock Vib.* 2021 (2021) e6363571. https://doi.org/10.1155/2021/6363571.

[4] *Libro Arquitectura Organica Memoria De La Exposicion En El Museo Nacional De Arquitectura, Javier Senosiain,* ISBN 9786074374148. Comprar en Buscalibre/*Book Organic Architecture Memory of the Exhibition at the National Museum of Architecture, Javier Senosiain,* ISBN 9786074374148. Buy at Buscalibre, (n.d.). https://www.buscalibre.us/libro-arquitectura-organica-memoria-de-la-exposicion-en-el-museo-nacional-de-arquitectura/9786074374148/p/49102839 (accessed March 14, 2022).

[5] Abdelaal, F., B.H.W. Guo, Stakeholders' perspectives on BIM and LCA for green buildings, *J. Build. Eng.* 48 (2022) 103931. https://doi.org/10.1016/j.jobe.2021.103931.

[6] Ikudayisi, A.E., A.P.C. Chan, A. Darko, O.B. Adegun, Integrated design process of green building projects: A review towards assessment metrics and conceptual framework, *J. Build. Eng.* 50 (2022) 104180. https://doi.org/10.1016/j.jobe.2022.104180.

[7] Avila, M., J. Toledo, F. Córdova, D. Icaza, M. de Los Angeles Tello, Intelligent Multifunctional Solar Urban Furniture: A multidisciplinary methodological vision of technology, in: *2018 Int. Conf. Smart Grid IcSmartGrid,* 2018: pp. 184–194. https://doi.org/10.1109/ISGWCP.2018.8634474.

[8] Antsaklis, P.J., J.A. Stiver, M. Lemmon, Hybrid system modeling and autonomous control systems, in: R.L. Grossman, A. Nerode, A.P. Ravn, H. Rischel (Eds.), *Hybrid Syst.,* Springer, Berlin, Heidelberg, 1993: pp. 366–392. https://doi.org/10.1007/3-540-57318-6_37.

[9] Yan, X., S.J. Wang, Infrastructure-Integrated Photovoltaic (IIPV): a boost to solar energy's green credentials? *Energy Procedia.* (2019). https://doi.org/10.1016/J.EGYPRO.2019.01.973.

[10] Fernández, M.C., M.F. Fernández, R.D. Fuentes, Á.C. Montiel, Calidad de la energía y generación distribuida en Cuba [Power quality and distributed generation in Cuba], *Rev. Cuba. Ing.* 1 (2010) 41–50. https://rci.cujae.edu.cu/index.php/rci/article/view/28 (accessed March 14, 2022).

[11] Deb, K., A. Pratap, S. Agarwal, T. Meyarivan, A fast and elitist multiobjective genetic algorithm: NSGA-II, *IEEE Trans. Evol. Comput.* 6 (2002) 182–197. https://doi.org/10.1109/4235.996017.

[12] Borge-Diez, D., J.M. Godoy-Déniz, Á. López-Rey, A. Colmenar-Santos, Pico turbines, the solution to self-supply energy to the water supply network. A case study in Las Palmas de Gran Canaria, *Energy.* 229 (2021) 120653. https://doi.org/10.1016/j.energy.2021.120653.

[13] Babatunde, O.M., J.L. Munda, Y. Hamam, How can a low-income household procure a small-scale hybrid renewable energy system? A techno-economic and acquisition analysis, *Int. J. Energy Sect. Manag.* 13 (2019) 1149–1172. https://doi.org/10.1108/IJESM-03-2019-0003.

[14] Hansen, K., B.V. Mathiesen, I.R. Skov, Full energy system transition towards 100% renewable energy in Germany in 2050, *Renew. Sustain. Energy Rev.* 102 (2019) 1–13. https://doi.org/10.1016/j.rser.2018.11.038.

[15] Goepfert, M., R. Castro, Techno-economic optimization of a 100% renewable energy system in 2050 for countries with high shares of hydropower: The case of Portugal, *Renew. Energy.* 165 (2021) 491–503. https://doi.org/10.1016/j.renene.2020.11.061.

[16] Cossent, R., T. Gómez, L. Olmos, Large-scale integration of renewable and distributed generation of electricity in Spain: Current situation and future needs, Energy Policy. 39 (2011) 8078–8087. https://doi.org/10.1016/j.enpol.2011.09.069.

[17] Krakowski, V., E. Assoumou, V. Mazauric, N. Maïzi, Reprint of Feasible path toward 40–100% renewable energy shares for power supply in France by 2050: A prospective analysis, *Appl. Energy.* 184 (2016) 1529–1550. https://doi.org/10.1016/j.apenergy.2016.11.003.

[18] EnergyPLAN, *Get Started,* (n.d.). https://www.energyplan.eu/getstarted/ (accessed November 1, 2021).

[19] de Montecristi, A.C. *Constitución de la República del Ecuador/*Constitution of the Republic of Ecuador, (2008). http://biblioteca.defensoria.gob.ec/handle/37000/823 (accessed December 15, 2021).

[20] Ayala, M., J. Maldonado, E. Paccha, C. Riba, Wind Power Resource Assessment in Complex Terrain: Villonaco Case-study Using Computational Fluid Dynamics Analysis, *Energy Procedia.* 107 (2017) 41–48. https://doi.org/10.1016/j.egypro.2016.12.127.

[21] Apolo, H., H. Apolo, K. Escobar-Segovia, D. Arcentales-Bastidas, Santa Cruz, *Galapagos Electricity sector towards a zero fossil fuel island,* (2019). http://laccei.org/LACCEI2019-MontegoBay/meta/FP170.html (accessed August 22, 2021).

[22] *80% de avance presenta Minas de Huascachaca, el proyecto eólico más grande del Ecuador* – Ministerio de Energía y Recursos Naturales no Renovables/*80% progress presents Minas de Huascachaca, the largest wind project in Ecuador –* Ministry of Energy and Non-Renewable Natural Resources, (n.d.). https://www.recursosyenergia.gob.ec/80-de-avance-presenta-minas-de-huascachaca-el-proyecto-eolico-mas-grande-del-ecuador/(accessed January 18, 2022).

[23] Arroyo M.F.R., L.J. Miguel, The Role of Renewable Energies for the Sustainable Energy Governance and Environmental Policies for the Mitigation of Climate Change in Ecuador, *Energies.* 13 (2020) 3883. https://doi.org/10.3390/en13153883.

[24] Arango, S., I. Dyner, E.R. Larsen, Lessons from deregulation: Understanding electricity markets in South America, *Util. Policy.* 14 (2006) 196–207. https://doi.org/10.1016/j.jup.2006.02.001.

[25] Mejía-Nova, F., D. Icaza, L. Cárdenas-Herrera, S.P. Galindo, C. Mejía-Cárdenas, Organic Constructions and Airplane-type Hostels in Isolated Places Supplied with Solar Energy: Case Study Arequipa-Peru, in: *2020 8th Int. Conf. Smart Grid IcSmartGrid,* 2020: pp. 48–55. https://doi.org/10.1109/icSmartGrid49881.2020.9144870.

[26] Icaza, D., D. Borge-Diez, S. Pulla Galindo, C. Flores-Vázquez, Modeling and Simulation of a Hybrid System of Solar Panels and Wind Turbines for the Supply of Autonomous Electrical Energy to Organic Architectures, *Energies.* 13 (2020) 4649. https://doi.org/10.3390/en13184649.

[27] Icaza, D., P. Chacho, D. Ramirez, J. Peñaloza, H. Mejía, S. Romero, Telecommunication System for the Monitoring of Tourist Hotels Type Airplane and Heritage Monuments Supplied by Solar Energy., in: *2019 IEEE Chil. Conf. Electr. Electron. Eng. Inf. Commun. Technol. CHILE CON,* 2019: pp. 1–8. https://doi.org/10.1109/CHILECON47746.2019.8987456.

[28] Icaza, D., C. Flores-Vázquez, J.-C. Cobos-Torres, S.P. Galindo, Organic Constructions and Airplane Type Hostels in Isolated Places Supplied with Solar Energy, in: M. Botto-Tobar, M. Zambrano Vizuete, P. Torres-Carrión, S. Montes León, G. Pizarro Vásquez, B. Durakovic (Eds.), *Appl. Technol.,* Springer International Publishing, Cham, 2020: pp. 231–243. https://doi.org/10.1007/978-3-030-42531-9_19.

[29] Dehghan Shabani, Z., R. Shahnazi, Energy consumption, carbon dioxide emissions, information and communications technology, and gross domestic product in Iranian economic sectors: A panel causality analysis, *Energy.* 169 (2019) 1064–1078. https://doi.org/10.1016/j.energy.2018.11.062.

[30] Almeida, R.M., A.S. Fleischmann, J.P.F. Brêda, D.S. Cardoso, H. Angarita, W. Collischonn, B. Forsberg, R. García-Villacorta, S.K. Hamilton, P.M. Hannam, R. Paiva, N.L. Poff, S.A. Sethi, Q. Shi, C.P. Gomes, A.S. Flecker, Climate change may impair electricity generation and economic viability of future Amazon hydropower, *Glob. Environ. Change.* 71 (2021) 102383. https://doi.org/10.1016/j.gloenvcha.2021.102383.

[31] World Energy Investment 2021 – Analysis, *IEA.* (n.d.). https://www.iea.org/reports/world-energy-investment-2021 (accessed July 12, 2021).

[32] Zhao, J.-T., B.-D. Su, S.K. Mondal, Y.-J. Wang, H. Tao, T. Jiang, Population exposure to precipitation extremes in the Indus River Basin at 1.5 °C, 2.0 °C and 3.0 °C warming levels, *Adv. Clim. Change Res.* 12 (2021) 199–209. https://doi.org/10.1016/j.accre.2021.03.005.

[33] Ayaburi, J., M. Bazilian, J. Kincer, T. Moss, Measuring "Reasonably Reliable" access to electricity services, *Electr. J.* 33 (2020) 106828. https://doi.org/10.1016/j.tej.2020.106828.

[34] Ratnam, E.L., K.G.H. Baldwin, P. Mancarella, M. Howden, L. Seebeck, Electricity system resilience in a world of increased climate change and cybersecurity risk, *Electr. J.* 33 (2020) 106833. https://doi.org/10.1016/j.tej.2020.106833.

[35] IRENA, Renewable Energy Policies in a Time of Transition: Heating and Cooling, *Publ.-Energy-Policies--Time--Transit.-Heat.--Cool.* (n.d.)./publications/2020/Nov/Renewable-Energy-Policies-in-a-Time-of-Transition-Heating-and-Cooling (accessed April 25, 2021).

[36] Shukla, K., W. Leal Filho, S. Lardjane, J.H. Sommer, A.L. Salvia, C. Borgemeister, The contribution of Regional Centers of Expertise for the implementation of the 2030 Agenda for Sustainable Development, *J. Clean. Prod.* 237 (2019) 117809. https://doi.org/10.1016/j.jclepro.2019.117809.

[37] Long-term production technology mix of alternative fuels for road transport: A focus on Spain, *Energy Convers. Manag.* 226 (2020) 113498. https://doi.org/10.1016/j.enconman.2020.113498.

[38] Idae, *SPAHOUSEC II: Statistical analysis of natural gas consumption in main dwellings with individual heating,* (n.d.). https://www.idae.es/publicaciones/spahousec-ii-analisis-estadistico-del-consumo-de-gas-natural-en-las-viviendas (accessed April 25, 2021).

[39] Gkillas, K., R. Gupta, C. Pierdzioch, Forecasting realized oil-price volatility: The role of financial stress and asymmetric loss, *J. Int. Money Finance.* 104 (2020) 102137. https://doi.org/10.1016/j.jimonfin.2020.102137.

[40] Blakers, A., M. Stocks, B. Lu, C. Cheng, The observed cost of high penetration solar and wind electricity, *Energy.* 233 (2021) 121150. https://doi.org/10.1016/j.energy.2021.121150.

[41] Martínez, J., J. Martí-Herrero, S. Villacís, A.J. Riofrio, D. Vaca, Analysis of energy, CO2 emissions and economy of the technological migration for clean cooking in Ecuador, *Energy Policy.* 107 (2017) 182–187. https://doi.org/10.1016/j.enpol.2017.04.033.

[42] Wu, Y., L. Zhang, Can the development of electric vehicles reduce the emission of air pollutants and greenhouse gases in developing countries*? Transp. Res. Part Transp. Environ.* 51 (2017) 129–145. https://doi.org/10.1016/j.trd.2016.12.007.

[43] De Luca, G., S. Fabozzi, N. Massarotti, L. Vanoli, A renewable energy system for a nearly zero greenhouse city: Case study of a small city in southern Italy, *Energy.* 143 (2018) 347–362. https://doi.org/10.1016/j.energy.2017.07.004.

[44] Bhore, S.J. Paris Agreement on Climate Change: A Booster to Enable Sustainable Global Development and Beyond, *Int. J. Environ. Res. Public. Health.* 13 (2016) 1134. https://doi.org/10.3390/ijerph13111134.

[45] Irigoyen Tineo, A. *Energy management system benchmarking for a remote microgrid*, (2017). https://e-archivo.uc3m.es/handle/10016/27629 (accessed August 22, 2021).

[46] Stytsenko, V.D., A.A. Lavarinenko, L.E. Nadra, V.A. Vinokurov, Prospects for improving the environmental and performance properties of motor fuels, *Chem. Technol. Fuels Oils.* 44 (2008) 364–369. https://doi.org/10.1007/s10553-008-0059-5.

[47] Bicalho, T., C. Bessou, S.A. Pacca, Land use change within EU sustainability criteria for biofuels: The case of oil palm expansion in the Brazilian Amazon, *Renew. Energy.* 89 (2016) 588–597. https://doi.org/10.1016/j.renene.2015.12.017.

[48] admin, Joint FAO-IRENA Report on Renewable Energy for Agri-food Systems Launched at COP26, *GSTDTAP.* (2021). http://119.78.100.173/C666//handle/2XK7JSWQ/340659 (accessed January 6, 2022).

[49] Folly, K.A., P.K. Olulope, G.K. Venayagamoorthy, Critical clearing time prediction using recurrent neural networks, in: *2017 Int. Jt. Conf. Neural Netw. IJCNN,* 2017: pp. 3303–3309. https://doi.org/10.1109/IJCNN.2017.7966270.

[50] Ordóñez, A., M. Daniela, Elección de tecnologías de energía renovable para uso en ciudades intermedias, utilizando criterios de sostenibilidad, caso de estudio, Cuenca-Ecuador/Choice of renewable energy technologies for use in intermediate cities, using sustainability criteria, case study, Cuenca-Ecuador, (2017). http://dspace.ups.edu.ec/handle/123456789/14762 (accessed March 5, 2021).

[51] Blake, M.B., H. Lund, The effectiveness of storage and relocation options in renewable energy systems, *Renew. Energy.* 33 (2008) 1499–1507. https://doi.org/10.1016/j.renene.2007.09.001.

[52] Braeuer, F., J. Rominger, R. McKenna, W. Fichtner, Battery storage systems: An economic model-based analysis of parallel revenue streams and general implications for industry, *Appl. Energy.* 239 (2019) 1424–1440. https://doi.org/10.1016/j.apenergy.2019.01.050.

[53] Adedoyin, F., I. Abubakar, F.V. Bekun, S.A. Sarkodie, Generation of energy and environmental-economic growth consequences: Is there any difference across transition economies? *Energy Rep.* 6 (2020) 1418–1427. https://doi.org/10.1016/j.egyr.2020.05.026.

[54] Alvarado, R., Q. Deng, B. Tillaguango, P. Méndez, D. Bravo, J. Chamba, M. Alvarado-Lopez, M. Ahmad, Do economic development and human capital decrease non-renewable energy consumption? E*vidence for OECD countries, energy.* 215 (2021) 119147. https://doi.org/10.1016/j.energy.2020.119147.

[55] Ahmadi, E., B. McLellan, T. Tezuka, The economic synergies of modelling the renewable energy-water nexus towards sustainability, *Renew. Energy.* 162 (2020) 1347–1366. https://doi.org/10.1016/j.renene.2020.08.059.

[56] Elliston, B., I. MacGill, M. Diesendorf, Comparing least-cost scenarios for 100% renewable electricity with low emission fossil fuel scenarios in the Australian National Electricity Market, *Renew. Energy.* 66 (2014) 196–204. https://doi.org/10.1016/j.renene.2013.12.010.

[57] Herremans, I.M., A.M. Echegaray, Biofuels in the Energy Mix of the Galapagos Islands, in: M.-E. Tyler (Ed.), *Sustain. Energy Mix Fragile Environ. Frame. Perspect.,* Springer International Publishing, Cham, 2018: pp. 49–56. https://doi.org/10.1007/978-3-319-69399-6_4.

[58] Bhatnagar, R., A. Gianoli, Water-for-energy in a climate change context - lessons learnt from Cuenca, Ecuador, Better World Actions Commit. *Sustain. Dev. Goals.* (2019) 43–45. https://research.tue.nl/en/publications/water-for-energy-in-a-climate-change-context-lessons-learnt-from- (accessed March 5, 2021).

[59] Goldemberg, J., R. Schaeffer, A. Szklo, R. Lucchesi, Oil and natural gas prospects in South America: Can the petroleum industry pave the way for renewables in Brazil? *Energy Policy.* 64 (2014) 58–70. https://doi.org/10.1016/j.enpol.2013.05.064.

[60] Perry, K.K. For politics, people, or the planet? The political economy of fossil fuel reform, energy dependence and climate policy in Haiti, *Energy Res. Soc. Sci.* 63 (2020) 101397. https://doi.org/10.1016/j.erss.2019.101397.

[61] Bogdanov, D., C. Breyer, North-East Asian Super Grid for 100% renewable energy supply: Optimal mix of energy technologies for electricity, gas and heat supply options, *Energy Convers. Manag.* 112 (2016) 176–190. https://doi.org/10.1016/j.enconman.2016.01.019.

[62] Nakazato, R.K., M.C.S. Rinaldi, M. Domingos, Will technological modernization for power generation at an oil refinery diminish the risks from air pollution to the Atlantic Rainforest in Cubatão, SE Brazil? *Environ. Pollut.* 196 (2015) 489–496. https://doi.org/10.1016/j.envpol.2014.05.011.

[63] Gould, C.F., S.B.S. Schlesinger, E. Molina, M.L. Bejarano, Household fuel mixes in peri-urban and rural Ecuador: Explaining the context of LPG, patterns of continued firewood use, and the challenges of induction cooking, *Energy Policy.* 136 (2020) 111053. https://doi.org/10.1016/j.enpol.2019.111053.

[64] La energía nuclear se mantiene como primera fuente de producción de electricidad en España pese al impulso de las renovables/Nuclear energy remains the leading source of electricity production in Spain despite the boost from renewables, LaSexta. (2021). https://www.lasexta.com/noticias/medio-ambiente/la-energia-nuclear-se-mantiene-como-primera-fuente-de-produccion-de-electricidad-en-espana-pese-al-impulso-de-las-renovables_202103286060fc8cdbb78d0001d4c48e.html (accessed June 24, 2021).

[65] Alves, D.B., F. Pérez-Cabello, Multiple remote sensing data sources to assess Spatio-temporal patterns of fire incidence over Campos Amazônicos Savanna Vegetation Enclave (Brazilian Amazon), *Sci. Total Environ. 601*–602 (2017) 142–158. https://doi.org/10.1016/j.scitotenv.2017.05.194.

[66] Tushar, M.H.K., A.W. Zeineddine, C. Assi, Demand-Side Management by Regulating Charging and Discharging of the EV, ESS, and Utilizing Renewable Energy, *IEEE Trans. Ind. Inform.* 14 (2018) 117–126. https://doi.org/10.1109/TII.2017.2755465.

[67] Secretaría Nacional de Planificación/National Planning Secretariat – Ecuador, (n.d.). https://www.planificacion.gob.ec/(accessed March 15, 2022).

[68] Agbanike, T.F., C. Nwani, U.I. Uwazie, L.I. Anochiwa, T.-G.C. Onoja, I.O. Ogbonnaya, Oil price, energy consumption and carbon dioxide (CO2) emissions: insight into sustainability challenges in Venezuela, *Lat. Am. Econ. Rev.* 28 (2019) 8. https://doi.org/10.1186/s40503-019-0070-8.

[69] Baumeister, C., L. Kilian, Forty Years of Oil Price Fluctuations: Why the Price of Oil May Still Surprise Us, *J. Econ. Perspect.* 30 (2016) 139–160. https://doi.org/10.1257/jep.30.1.139.

[70] Abril, M.Á.C. Assessment of Adaptation Measures to Climate Change in The Paute River Watershed and Its Hydropower Plants, *Graduate Studies,* 2011. http://dx.doi.org/10.11575/PRISM/35850.

[71] Castro Verdezoto, P.L., J.A. Vidoza, W.L.R. Gallo, Analysis and projection of energy consumption in Ecuador: Energy efficiency policies in the transportation sector, *Energy Policy.* 134 (2019) 110948. https://doi.org/10.1016/j.enpol.2019.110948.

[72] Gregor Barié, C. Nuevas narrativas constitucionales en Bolivia y Ecuador: el buen vivir y los derechos de la naturaleza, Latinoam/New constitutional narratives in Bolivia and Ecuador: good living and the rights of nature, Latin America, *Rev. Estud. Latinoam.* (2014) 9–40. http://www.scielo.org.mx/scielo.php?script=sci_abstract&pid=S1665-85742014000200002&lng=es&nrm=iso&tlng=es (accessed August 29, 2020).

[73] Villalba-Eguiluz, C.U., I. Etxano, Buen Vivir vs. Development (II): The Limits of (Neo-) *Extractivism, Ecol. Econ.* 138 (2017) 1–11. https://doi.org/10.1016/j.ecolecon.2017.03.010.

[74] Brugger, H., W. Eichhammer, N. Mikova, E. Dönitz, Energy Efficiency Vision 2050: How will new societal trends influence future energy demand in the European countries? *Energy Policy.* 152 (2021) 112216. https://doi.org/10.1016/j.enpol.2021.112216.

[75] *Net-zero by 2050 hinges on a global push to increase energy efficiency – Analysis,* IEA. (n.d.). https://www.iea.org/articles/net-zero-by-2050-hinges-on-a-global-push-to-increase-energy-efficiency (accessed July 14, 2021).

[76] Abbas, A.O., B.H. Chowdhury, Using customer-side resources for market-based transmission and distribution level grid services – A review, *Int. J. Electr. Power Energy Syst.* 125 (2021) 106480. https://doi.org/10.1016/j.ijepes.2020.106480.

[77] El gobierno de Ecuador deroga el decreto que eliminaba el subsidio a los combustibles y se anuncia el fin de las protestas [The government of Ecuador repeals the decree that eliminated the fuel subsidy and the end of the protests is announced], *BBC News Mundo.* (n.d.). https://www.bbc.com/mundo/noticias-america-latina-50037781 (accessed August 27, 2021).

[78] Alvarez, D.I. Modeling, simulation and construction of the D-ICAZA-A1 wind turbine destined for the rural areas of Ecuador, in: *2017 IEEE PES Innov. Smart Grid Technol. Conf. - Lat. Am. ISGT Lat. Am.,* 2017: pp. 1–6. https://doi.org/10.1109/ISGT-LA.2017.8126742.

[79] Icaza, D., H. Conce, P. Flores, F. Conce, Dimensioning of the main mechanical elements and final assembly of the DIAWIND-A2 wind turbine, in: 2017 Chil. *Conf. Electr. Electron. Eng. Inf. Commun. Technol. CHILE CON*, 2017: pp. 1–6. https://doi.org/10.1109/CHILECON.2017.8229551.

[80] Kuo, B.C., G. Aranda Pérez, *Sistemas de control automático/Automatic control systems*, 7a ed, Prentice-Hall Hispanoamericana, México, 1996.

[81] Ogata, K., R. Dormido Canto, S. Dormido Canto, *Ingenieria de control moderna/Modern control engineering*, 4a ed, Prentice Hall, Madrid, 2005.

[82] Brito, P., J. Manuel, *Análisis de factibilidad para la implementación de sistemas híbridos (Eólico-fotovoltaico) en las inmediaciones del centro patrimonial de Quingeo y zona arqueológica Curiquinga/Feasibility analysis for the implementation of hybrid systems (Wind-photovoltaic) in the vicinity of the Quingeo heritage center and the Curiquinga archaeological zone* (2018). http://dspace.ucacue.edu.ec//handle/reducacue/8130 (accessed November 3, 2020).

Chapter 6

Wind Distributed Generation and Desalination: An Application to a Rural Microgrid

David Borge-Diez[1], Francisco José García-Moya[1] and Enrique Rosales-Asensio[2,*]
[1]Departament of Electrical, Systems and Automation Engineering, University of León, León, Spain
[2]Department of Electrical Engineering, University of Las Palmas de Gran Canaria, Las Palmas de Gran Canaria, Canary Islands, Spain

Abstract

In both isolated and non isolated systems - from food to fossil-fuel sources - all the exogenous inputs must be imported, although isolated ones comprise higher economic and environmental costs. Likewise, indigenous water sources on water-scarce islands are generally unable to meet the needs of inhabitants, livestock feed or agriculture. This fact has provoked the development of different water desalination technologies. Among these, reverse osmosis stands out as the world's most widespread technique.

Wind energy is one of the renewable energy resources that has grown the most during the last decade. However, the intermittent nature of wind energy, and especially the variable speed of wind are some of the drawbacks of this generation system.

A common solution to mitigate intermittency and increase system availability consists in using wind generation in parallel with other

*Corresponding Author's Email: enrique.rosales@ulpgc.es.

In: Wind Turbines
Editors: David Borge-Diez and Daniel Icaza
ISBN: 978-1-68507-974-1
© 2022 Nova Science Publishers, Inc.

renewable energy sources and storage means such as pumped hydroelectric storage, one of the most economically viable methods. Among the challenges of integrating wind and pumped hydroelectric storage, the following must be highlighted: optimizing energy consumption, production, and storage.

As long as reversible turbine/generator assemblies can act as pumps or turbines, renewable energy can be used to develop a hybrid desalination plant that provides water as well as energy to a microgrid of renewable energy sources. When water and energy are managed in an integrative manner, the dual system boosts the synergies between the water-energy-food systems, and the efficiency, safety and reliability of the hybrid system increase. The system here proposed increases the efficiency and reliability of desalination plants, as well as the security of electricity supply in a microgrid.

Keywords: electrical energy hub, desalination wind, hybrid management system, water-energy nexus

Introduction

One of the major challenges that humankind faces in this century is how to achieve welfare while maintaining environmental sustainability in an accelerating global environmental threat. It highlights feeding the growing world population as one of the most challenging activities. Models from United Nations (UN) forecast that the world population will reach 9.7 billion by 2050 [1], leading to an increasing global food production requirement. Among the major climate change concerns, it highlights the CO_2 emissions. As can be seen in Figure 1, emissions regarding human activity are constantly raising. It highlights electricity and heat producers, transport or industry as the major world CO_2 emissions contributors. In addition to the growing vegetables, or cattle feeding CO_2 impacts, it must also incorporate all agri-food value chain. This includes processing industry, food preservation with refrigeration techniques, and all trade and transport around activity, Figure 1.

It's a well-known fact that islands suffer from isolation issues. Often relying on fossil fuel-based power plants, it used to be necessary to import high contaminating fossil fuel-based energy sources provoking higher costs. As far as the water supply is concerned, increasing water scarcity risks results in extra demands on the water supply system. The desalination industry supplies water for drinking and agriculture in the world's arid coastal regions.

Among the most used techniques for water desalination, it highlights Reverse Osmosis (RO), Multistage flash evaporation (MEF), or electrodialysis (ED). Reverse osmosis (RO) is the leading desalination technology with 39.4 million m^3 of global installed capacity [3]. Independently of the water desalination technology, energy (thermal or electrical) accounts for around 60-75% of total costs. Different studies have assessed water desalination costs, highlighting energy, operating and maintenance costs [4]. Escobar et al. [5] estimated the total water costs ($\$/m^3$) in different plants. According to Arenas et al. [6], current desalination technology makes it possible to desalinate a cubic meter of seawater for less than 2 kWh on a medium or large scale, considering only the desalination process.

Figure 2 shows the environmental consequences for a system of both desalination and isolation. As water demand increases, desalinated water demand increases, as well as costs and related emissions. Since the energy used in the desalination process comes from fossil fuels, associated climate change risks increase as well. In addition, this leads to a lower rainfall scenario provoking increasing water demand, and increasing desalted water needs. Specifically, in water scarce islands, indigenous water sources cannot supply inhabitants, agriculture, and cattle food. In these, water desalination becomes essential to provide water. Isolation leads to greater distance and fewer interconnection modes for importing goods.

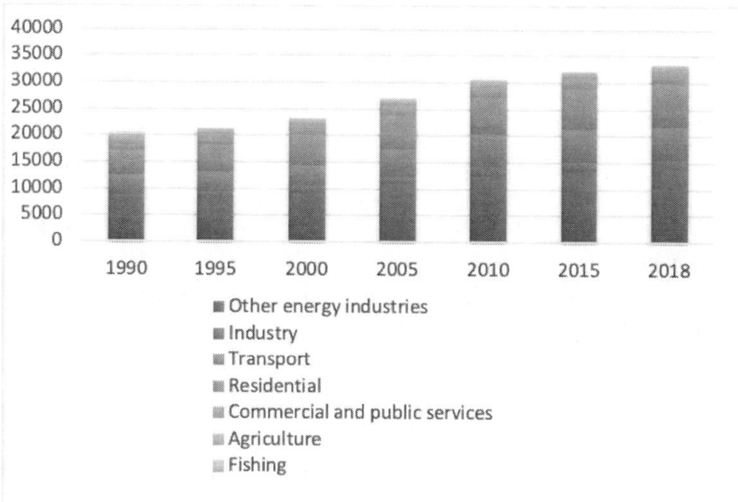

Source: [2].

Figure 1. CO_2 emissions (Mt of CO_2) by source.

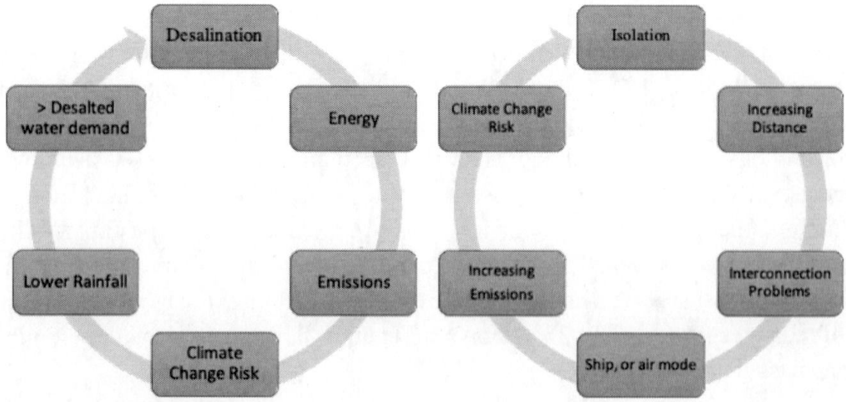

Source: own elaboration.

Figure 2. Environmental consequences in water scarce islands.

Hence, the isolation from the mainland system provokes that goods only reach the island by sea or air. Due to an island's isolated nature, many goods must be ship imported, including energy sources and food. Annually, international shipping is estimated to produce around 18.6 million and 10.6 million tons of NOx (as NO_2) and SOx (as SO_2). According to IPPC (Intergovernmental Panel on Climate Change) [7] global NOx and SOx emissions from all shipping represent about 15% and 13% of global NOx and SOx from anthropogenic sources. CO_2 emissions from shipping are approximately 850 million tons per year.

Renewable Energy Deployment in Islands

It's a well-known fact that electricity is supposed to play an important role to reduce greenhouse gas emissions, and other pollutants that energy consumption provoke [8]. In the last 10 years, renewable-energy generation technologies falling prices provoked an increasing share of renewable electricity among the power generation systems. Technologies such as solar photovoltaic or wind have lowered their Levelized Cost of Electricity (LCOE) [9, 10]. This has led to an increased economic competitiveness versus fossil fuel energy sources. Climate change concerns, pressure over fossil fuel energy sources, and others boost their deployment as well. Especially wind energy [11] deployment has increased over the last few years. Global annual average increase in wind energy production in 2018-2019 is 12%, and it is forecasted that both on-shore and off-shore will continue to expand until 2025 [12].

There are several initiatives have been taken to boost the renewable energy share in the generation mix of isolated islands. Among these are projects for increasing the wind/solar capabilities in islands, such as O_2 projects [13], which include projects in Atlantic, Indian Ocean, Mediterranean, and South China sea. The Hawaii Clean Energy Initiative (HCEI) for the state of Hawaii [14] included a mandate to generate 40% of its energy from renewable resources by 2030 in Oahu Island. Increasing renewable energy capacity in isolated systems is a common method to reduce environmental, and under certain circumstances, also reduce monetary costs. However, the intermittent nature and the variable speed of wind or cloud density are important drawbacks of renewable-energy generation systems. These concerns are more pronounced in islands due to the isolated nature that makes it difficult to share or import energy. In addition, the system faces faster and more severe frequency deviations in cases of sudden changes in supply or demand [15], increasing system costs.

Among the most sustainable energy storage means, it highlights Pumping Hydroelectric Storage (PHS), but others like Hydrogen, or superconducting magnetic energy storage are expected to increase. Hydroelectric pumping storage is highlighted as one of the most economically and viable supporting methods. Among the benefits: high efficiency, water turbine constant start-up costs that reduce environmental costs versus thermal units. According to Pérez-Díaz et al. [16] in an isolated system with high wind power penetration, the costs of power systems might be reduced, just managing the energy surpluses with pumping hydroelectric storage means. Some studies have been developed for assessing, sizing, and operating pumping hydroelectric storage means in islands. Katsaprakakis et al. proposed a scheme that uses sea water [17]. Portero et al. focused in the Gran Canaria's system [18]. Focusing on the economic behavior of facilities in a system is especially important for evaluating their operating performance. Through stochastic optimization, Rosales-Asensio et al. [19] sized an economic profitable wind powered desalination plant. In addition, Borge et al. [20] investigated the amount of water and energy to be sold for an economically-profitable wind-powered desalination plant under a grant scenario.

A Step towards Sustainability in Water Scarce Areas

Achieving food security is one of the agenda 2030's [21] major objectives. This becomes an enabler for sustainable development in its three dimensions: economic, social, and environmental. According to Alexandratos [22] it is needed an increase of 70-100% in the food supply chain for maintaining current nutrition levels. Agri-food industry is the world's largest consumer of freshwater, accounting for 70% of total global water withdrawals [23]. In addition, around 30% of the global energy consumption occurs along food supply chains [24]. As previously mentioned, this fact leads to a correlated impact in the water resource as well in the CO_2 associated emissions that must be taken into account for the food industry.

At the same time, agriculture is the most climate-vulnerable sector, subject to impacts such as those provoked for longer water scarcity periods, extreme weather events, as well as region-specific changes in crop productivity. According to Schneider et al. [25], energy intensity in agriculture increased until the 1980's, lowering during the following decades. Hernandez et al. [26] classified regions regarding the Standardized Precipitation Index (SPI), and found for a high-emission scenario that water scarce regions will increase during next decades. McKee et al. developed the SPI index for assessing precipitation deficit on any time scale [27]. WEF (Water, energy, food) nexus analysis tools are used for the evaluation of system's sustainability [28]. Through different approaches and system's design, its aim is to measure trade-offs between systems for the sustainability evaluation of a system. The better understanding of the trade-offs leads to a more effective use of resources. Sustainability of a system may be evaluated by quantitative means. Rosales et al. [29] proposed a tool to measure a system deviation from an acceptable scenario.

Location Strategies

In order to provide renewable energy surpluses through sustainable management and storage through hydraulic pumping deployment in isolated islands, different strategies have been proposed. Portero et al. studied how to take advantage of the near to sea location for storing energy [30]. It is worth to note that renewable-powered RO plants standout as a solution for lowering their high energy costs. The hybrid plant takes advantage of renewable energy for developing a hybrid wind-solar energy self-consumption system. Either off-shore, or on-shore, wind energy not only provides energy to a microgrid. At the same time, the hybrid plant provides water for crop irrigation and for the inhabitants, thus reducing the energy consumption of the desalination plants, as well water consumption [31, 32]. Despite its higher costs versus onshore and solar PV, offshore wind deployment has benefits versus onshore projects including that offshore wind speeds are generally faster, and don't vary as much. In addition, wind direction doesn't change as often. This leads to increasing reliability, and energy.

Wind energy is often higher in islands, especially on the coast where there are no physical barriers. [33]. They not only have a smaller visual impact as onshore turbines, but they can also be built taller. It is expected an average growth of nearly 30% per year since 2010, and more than 100 new offshore wind projects are scheduled around the world [34]. On the other hand, operating and maintaining costs increase costs due to the distance to the array. The environmental impact of offshore wind deployment includesto installation, operation and maintenance - either preventive or corrective, and its future decommissioning. According to Mauricio Hernandez et al. [35] among the environmental costs is burying the offshore cables to protect them from fishing gear and ship anchors. It is also highlighted that it can impact fauna and marine life, and the environmental concerns due to the installation of the jacket on the seabed. The operation and maintenance objective are to verify that the key performance indicators (KPIs) of the offshore wind turbine project are achieved. Generally, maintenance can be divided into two parts: preventive, and corrective maintenance. Decommissioning is the stage for removing all the wind farm components, including among others the wind turbine, foundation, transition piece, cables, substation.

Energy Management Solutions in Hybrid Desalination Plants

Provided that renewable resource is available, from-the-grid energy consumption in the desalination plant will fall due to self-consumption. Additionally, its economic performance will improve, because of the increasing energy share in the island's generation system, boosted by lower resource prices. As renewable energy share increases, energy storage system needs also increase. Among the challenges for storing energy are: optimizing energy consumption and production, reliability, stability, quality, and storage maximizing. Irrespective of the storage method used, agile energy management systems for supporting the renewable-based energy generation system will be needed. Sustainable energy storage means are mandatory not only for maintaining system stability, but also for increasing the environmental benefits of renewable energy.

Among the proposed energy models there is a focus on the operational, or planning phase. Computer-design analysis tools boost the renewable energy integration in energy systems [36]. Energy Plan [37] uses energy demand, renewable energy sources, energy station capacities, costs and a number of optional regulation strategies as general inputs [38]. As outputs, this tool returns the energy balance and resulting annual, monthly and hourly energy productions, fuel consumptions, electricity import/export and total annual carbon dioxide emissions. Through constant improving, these tools facilitates getting an up to-date system for a better modelled system. Cabrera et al. [39] proposed a tool for easing the planning strategy. The renewable energy control system invention here proposed manages desalinated water through a coupled energy control system for renewable energy - wind and solar photovoltaic powered desalination plant. The system implements some optimization methods that benefit from the resource management. Figure 3 depicts a standard energy model with a suggested coupling to a fluid operation management network.

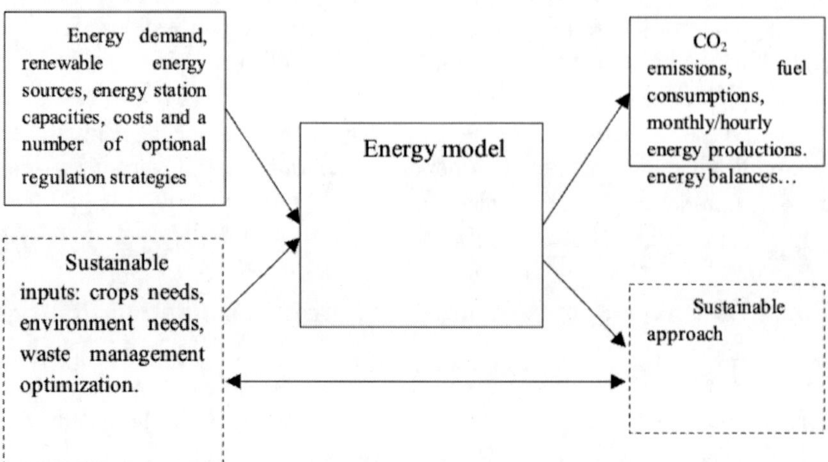

Figure 3. I/O approach for water, renewable energy, and food models.

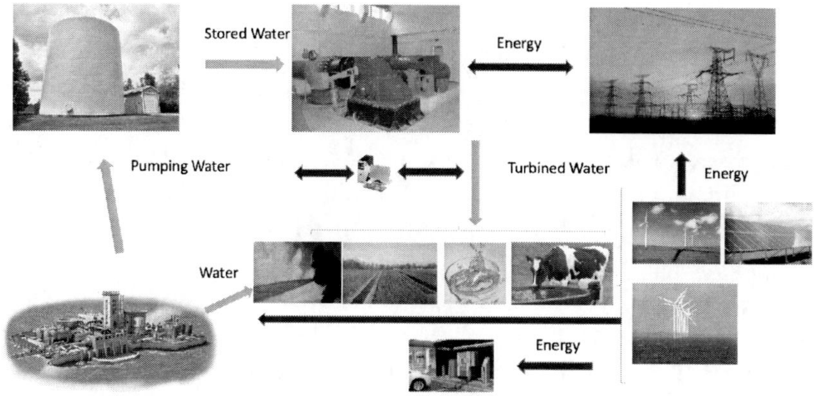

Source: Own elaboration.

Figure 4. Control system of a hybrid desalination plant.

Including a renewable powered, and grid connected desalination plant, Figure 4 represents the elements constituting a system that takes advantage of desalinated water to develop a pumping hydro-storage scheme. When operated in an integrative manner, the water-energy-food subsystems synergies improve.

The system consists in controlling the renewable-based energy generation system, the energy consumption in the desalination plant, as well of the PHS storage pumps. The system monitors the energy storage state of the water reservoirs, analyzes the renewable-powered desalination plant, as well the microgrid system energy needs to decide the most energy-sustainable choice. Only by energy generation, consumption, and storage dynamic management, the renewable energy share in the system could increase. As long as offshore wind energy benefits from higher wind energy density [35] it stands out as a solution for renewable-based desalination plants, as well for the water pumping hydro-storage mean. In order to reduce energy losses, it is needed coordinated operation of the renewable generation plant, the desalination plant that implements the energy storage pumping scheme, and energy needs in the system.

The proposed system shows how to optimize energy usage in the microgrid. In order to do this, it manages flow in a desalination plant, while matching renewable energy generation system with the energy needs in the facility, as well as the microgrid to which it is connected. Specifically, it uses electricity energy generation, and consumption information to create a hybrid renewable-powered microgrid. It balances the energy needs in the desalination plant, microgrid, as well in the energy storage pumping system, or in the water supply system. In order to maximize resource value, it satisfies operating conditions in the two-sided energy generation scheme. In addition, the solution to the technical problem must include the development of a computer implemented method that utilizes context data to maximize the operation of the desalination microgrid. This provides bidirectional control of the energy components, including recording energy utilization (consumption/production/ storage), data and directing operation of the energy components.

The technical problem according to the system intends to maximize the renewable energy generated in a microgrid. The environmental problem that the proposed model deals with relates to resource-use maximization. It pumps desalinated water for taking advantage of energy surpluses in grid-connected operating conditions. Specifically, it allows a wide operation range to be covered in grid-connected desalination plants. The system must compensate flow of energy and desalinated water to provide a solution for renewable energy management. It increases the sustainability of renewable energy - mainly offshore wind energy - through the integral system control. The system should, furthermore, provide an enforceable safety operating condition that allows focusing in operation needs, and in system sustainability. On the other hand, the technical problem that the system deals with relates to the inability of implementing a coordinated operation of the renewable generation plant, and the desalted water in the implemented energy pumping hydro-storage mean. The system manages water from desalination plants implementing a hydraulic pumping energy storage mean with input-output based on irrigation, and energy needs. Combined with a grid-connected self-consumption renewable based energy electricity generation system, the desalination plant and microgrid, control system enables dynamic management of energy consumption/production/storage.

The computer implemented method here proposed increases the reliability of desalination plants thanks to redundant power generation energy systems. These include energy from the grid, self-consumption, and the storage system. As long as there is more back-up energy, the plant's performance parameters will improve, and there will be fewer energy-related shutdowns. The economic performance of the facility increases provided that the energy surpluses are sold in the electricity market [20]. In another aspect, the optimization engine in the computer implemented system for energy management provides additional environmental benefits.

Some of the systems that intend to use and control pumped hydro energy storage systems combined with energy from renewable sources focus only on energy control. Navas et al. [40] proposed a wind-based pumped-storage supplementary electricity generation control system and method. Haibao [41] patented a system for lifting water through solar photovoltaic and making redundant electricity get on grid. Specifically, others proposed energy systems for desalination plants control. Peng et al. [42] designed a device to produce desalinated water that takes advantage of wind energy. Allen et al. [43] designed an integrated system for generating, storing and dispensing clean energy and desalted water. However, these only focus on the energy control of the facility, monitoring the consumed and generated energy, making decisions based on its management (e.g., pumping, turbine). These control systems for desalination plants that incorporate a pumping hydro-energy storage system powered with renewable energy only focused on energy control, monitoring the consumed and generated energy in order to decide to activate the pump or turbine on the basis of this information. These control systems do not include a sustainable management method for controlling the energy, neither the waste from productive process.

Wind Based System Description

The system becomes a solution in order to increase renewable energy share in the microgrid the desalination plant belongs to. As long as energy self-consumption in desalination plants increases, systems like the one here presented are essential not only to maintain the energy electricity quality parameters, but also to increase the electric system environmental sustainability. The system includes a grid-connected renewable-based energy electricity generation self-consumption system that powers the desalination plant. Specifically, it includes at least one wind turbine generator device configured to generate output power for a common bus; at least one hydro generator device configured to generate output power for said common bus, said at least one hydro generator device powered by desalinated water flow from upper reservoir; and said at least one hydro generator device having corresponding local controls associated therewith; and a set of supervisory controls in communication with said common bus and each of said local controls. The water pumping hydro energy storage system consists in two tanks at different heights, connected to the external system itself. After the water goes through the turbine, it is used for supplying water to the orchards, for firefighting, among others uses. The system might include intercalated mechanisms to place fire extinguishing hoses, such as quick hose connections to connect the fire-fighting equipment in order to fight fires. The water rising hoses are preferably placed at sight and in black, so that the fluid is heated, and with a controlled section reduction to facilitate water rising.

The pumping hydro storage system is designed to be placed in the desalination plant surroundings, in the side of a mountain, hill, boulevard, etc., taking advantage of its steep slope. In order to find the most ideal locations to place the upper water reservoirs of hybrid wind-powered desalination systems, GIS tools become essential. Borge Díaz et al. [44] designed a strategy to find the most promising locations to implement the PHS strategy in Gran Canaria RO desalination plants. As previously mentioned, building, O&M, and decommissioning jobs of renewable energy in general, and particularly of wind energy projects in rural areas increases the capacity for fixing rural population, especially in regions at high water scarcity risk. In the presented microgrid, wind energy boosts the development of a water, energy, and food self-sufficient study scenario, enabling the analysis of systems with sustainability indexes. The hybrid desalination plant provide water as well energy in a lack-of-resources environment. The presented system provides an opportunity to maximize synergies between the water-energy-food subsystems. The WEF nexus analysis investigate trade-offs between these elements, and across different levels and dimensions. Wind energy boost the benefits of the WEF nexus management strategy. From both energy-water, and energy-food synergies, renewable energy increases the system's sustainable management.

The wind powered system here presented allows reducing water consumption of the energy electricity generation sector, since wind energy is one of the less water consuming technologies [45]. It also allows to improve resource management strategies, especially in isolated, and water-scarce areas. These regions benefit from an easier access to seawater. As previously mentioned, redundant power systems increase reliability of desalination plants supply, lowering a bridge for closing the water gap in water scarce areas. According to Serrano-Tovar [46], the trade-offs across different levels and dimensions of analysis can be anticipated. It is increasingly important to decouple agri-food chain from fossil fuels. In order to afford sustainable food for the increasing world population, it is vitally important to reduce the associated CO_2 equivalent emissions. In addition, the system here presented provides a sustainable and reliable access to water supply for agriculture use. According to Châtel et al. [47] the Mediterranean region is one of the most water-scarce areas in the world. The MENA (Mediterranean and North Africa) region holds only 3% of the world's freshwater resources. It also hosts more than 50% of the world's "water poor" populations, encompasing 180 out of the 460 million inhabitants of the region. Economic activities are specifically harder in islands, and even more difficult in those suffering lack of water. In order to integrate planning strategies, WEF (Water, Energy, and Food) nexus analysis outstands as a method for an integrated environmental planning [48]. The ability to evaluate planning takes on greater relevance in regions where water scarcity becomes a pressing issue. Emerging economies grow rapidly by shifting toward secondary manufacturing, leading to improved human development indices, increased urbanization and development of economic and social infrastructure, often with considerable remaining rural poverty. The growing desalination industry produces drinking water, and water for agriculture in the world's arid coastal regions, but it leaves behind a waste product of highly concentrated brine which is dangerous for marine ecosystems. It is usually disposed of by dumping it back into the sea, a process that requires costly pumping systems and that must be managed carefully to prevent damage to environment. In order to increase a system's sustainability it is necessary to include the energy perspective for taking advantage of the waste (RO-hybrid desalination plant -brine) to manufacture high added-value products, or further transforming processes. As previously mentioned, exploiting water-energy-food eco-system synergies will be vitally important to improve renewable energy sustainable management efficiency. In order to achieve increasingly sustainable and efficient energy systems, the decision-making process must choose the most sustainable, and economically efficient option in the dual system. Only by boosting the positive synergies of the desalination plant, and the storage unit the system maximizes the operating parameters in the hybrid powered desalination plant. Among the positive effects, these systems are able to fix population in water scarce and rural areas.

Conclusion

In order to boost the sustainable development of drought-threatened eco-systems, management systems are mandatory to increase the positive synergies between the water, renewable energy, and food systems. As presented, it is also important to evaluate actions that might cause long-term environmental problems. Equally important is to decide between different renewable energy deployment strategies, as well as the economic performance of renewable-powered desalination plants. The system here presented intends a coordinated operation of the wind energy plant, the energy pumping hydro-storage system, and the desalination plant. In order to reduce water scarcity consequences, the proposed system for controlling renewable-powered desalination plants combined with the hydraulic pumping energy storage system is focused not only in monitoring energy, but also on sustainability concerns. On the basis of the information gathered, such as resource price, electricity, etc., it carries out a decision-making process to optimize energy management within a microgrid.

References

[1] United Nations. 2021. *"Population."* Accessed July 21. https://www.un.org/en/global-issues/population.

[2] International Energy Agency. 2021. *"Data and statistics."* Accessed July 21, https://www.iea.org/data-and-statistics/?country=WORLD&fuel=Energy%20 transition%20indicators&indicator=CO2BySector.

[3] osmosis (RO) is the desalination leading technology with 39.4 million m3.

[4] Wittholz, Michelle K., O'Neill, Brian K., Colby, Chris B., Lewis, David. 2008. "Estimating the cost of desalination plants using a cost database." *Desalination* 229:10-20.

[5] Escobar, I. C., Schäfer, A. I. 2010. *"Sustainable Water for the Future Water Recycling Versus Desalination."* Academic Press, Elsevier: Cambridge, Massachussets.

[6] Arenas Urrea, S., Dias Reyes, F., Peñate Suarez, B., de la Fuente Bencomo, J. A. 2019. "Technical review, evaluation and efficiency of energy recovery devices installed in the Canary Islands desalination plants." *Desalination*, 450:54-63.

[7] Pachauri, R. K., and Meyer, L. A. 2014. *"Climate Change 2014: Synthesis Report. Contribution of Working Groups I, II and III to the Fifth Assessment Report of the Intergovernmental Panel on Climate Change."* Geneva: IPCC.

[8] Díaz, A. R., Ramos-Real, F. J., Marrero, G. A., Perez, Y. 2015. "Impact of Electric Vehicles as Distributed Energy. Storage in Isolated Systems: The Case of Tenerife." *Sustainability* 7:15152-15178.

[9] International Renewable Energy Agency. 2021. *"Technologies."* Accessed July 22. https://www.irena.org/Statistics/View-Data-by-Topic/Capacity-and-Generation/Technologies.

[10] International Renewable Energy Agency. 2018. "Renewable Power Generation Costs in 2017," International Renewable Energy Agency, Abu Dhabi.

[11] Manwell J. F., McGowan J. G., Rogers A. L. 2009. *"Wind energy explained."* Great Britain: John Wiley & Sons.

[12] International Energy Agency. 2021. *"World Energy Balances: Overview."* Accessed August 5. https://www.iea.org/reports/world-energy-balances-overview.

[13] IRENA. *"A path to prosperity: renewable energy for islands,"* 2016.

[14] Summary Report Oahu Wind Integration and transmission study.

[15] Zhao, H., Wu, Q., Hu S., Xu H., Rasmussen, C. N. 2014. "Review of energy storage system for wind power integration support." *Applied Energy* 137:545-53. http://dx.doi.org/10.1016/j.apenergy.2014.04.103.

[16] Pérez-Díaz, J. I., and Jiménez, J. 2016. "Contribution of a pumped-storage hydropower plant to reduce the scheduling costs of an isolated power system with high wind power penetration." *Energy* 109.

[17] Katsaprakakis Dimitris Al, Christakis Dimitris G., Stefanakis, Ioannis, Spanos, Petros, Stefanakis, Nikos. 2013. "Technical details regarding the design, the construction and the operation of seawater pumped storage systems." *Energy* 55:619-630.

[18] Portero, U., Velázquez, S., and Carta, J. A. 2015. "Sizing of a wind-hydro system using a reversible hydraulic facility with seawater. A case study in the Canary Islands." *Energy Conversion and Management*, 106:1251-1263. https://doi.org/10.1016/j.enconman.2015.10.054.

[19] Rosales Asensio, E., and Colmenar Santos, A. E., 2021. "Surrogate optimization of coupled energy sources in a desalination microgrid based on solar PV and wind energy." *Desalination* 500. https://doi.org/10.1016/j.desal.2020.114882.

[20] Borge Diez, D., García Moya, F. J., Cabrera Santana, P., and Rosales Asensio, E. 2020. "Feasibility analysis of wind and solar powered desalination plants: an application to islands." *Science of the Total Environment*. https://doi.org/10.1016/j.scitotenv.2020.142878.

[21] United Nations. 2021. *"Transforming our world: the 2030 Agenda for Sustainable Development."* Accessed August 15. https://sdgs.un.org/2030agenda.

[22] Alexandratos, N., and Bruinsma J. 2012. "World Agriculture Towards 2030/2050: The 2012 Revision," *ESA Working Paper*, 12-03.

[23] Food and Agriculture Organization of the United Nations. 2011a. "*The state of the world's land and water resources for food and agriculture (SOLAW) - Managing systems at risk"*. Accessed August 15. http://www.fao.org/docrep/017/i1688e/i1688e.pdf.

[24] Food and Agriculture Organization of the United Nations. 2011b. *"Energy-smart food for people and climate.*" Issue Paper." Accessed august 23. http://www.fao.org/docrep/014/i2454e/i2454e00.pdf.

[25] Schneider, U. A., and Smith, P. 2009. "Energy Intensities and Greenhouse Gas Emissions in Global Agriculture." *Energy Efficiency* 2:195-206.

[26] Hernandez, Yeray, Barbosa, Paulo, Guimarães Pereira, Ângela, Palacios-Díaz, M del Pino, Hernandez, Sara, Bordón, Sebastián, Suárez, Baltasar, Vasseur, Ezequiel, Ruano, Juan, Monterrey-Viña, Adrián, Mendoza-Grimón, Vanessa, Martel, Gilberto, Gómez-Gotor, Antonio, Fernández-Pinazo, Fernando, Lozano-Medina, Alexis. 2020. *"Acequia report: climate change, droughts and water uses. A LOGIC proposal for action."* 10.2760/466906.

[27] McKee, T., Doesken, N., and Kleist, N., 1993. *"The relationship of drought frequency and duration to time scales."* Boston, American Meteorological Society, pp. 179-184.

[28] Nhamo, Luxon, Ndlela, Bekithemba, Nhemachena, Charles, Mabhaudhi, T., Mpandeli, Sylvester, Matchaya, Greenwell. 2018. "The Water-Energy-Food Nexus: Climate Risks and Opportunities in Southern Africa." *Water*. 10.

[29] Rosales-Asensio, Enrique, de la Puente-Gil, Álvaro, García Moya, Francisco José, Blanes Peiró, Jorge, de Simón-Martín, Miguel. 2020. "Decision-making tools for sustainable planning and conceptual framework for the energy-water-food nexus." *Energy Reports* 6:4:15. https://doi.org/10.1016/j.egyr.2020.08.020.

[30] Portero, U., Velázquez, S., and Carta, J. A. 2015. "Sizing of a wind-hydro system using a reversible hydraulic facility with seawater. A case study in the Canary Islands." *Energy Conversion and Management*. 106, 1251-1263. https://doi.org/10.1016/j.enconman.2015.10.054.

[31] Rosales Asensio, E., García Moya, F. J., González Martínez, A., Borge-Diaz, D., de Simón Martín, M., 2019. "Stress mitigation of conventional water resources in water-scarce areas through the use of renewable energy powered desalination plants: an application to the Canary Islands." *Energy Reports*. 6:124-135. https://doi.org/10.1016/j.egyr.2019.10.031.

[32] Lozano-Ruano, J. B., S. L. Soslaires Canarias. 2018. Accessed July, 8 http://www.islhagua.itccanarias.org/c/document_library/get_file?p_l_id=22263&folderId=22087&name=DLFE-925.pdf.

[33] Schallenberg Rodríguez, J., García Montesdeoca, N. 2018. "Spatial planning to estimate the offshore wind energy potential in coastal regions and islands. Practical case: the Canary Islands." *Energy* 143:91-103.

[34] IEA International Energy Agency. 2019. *"Offshore wind outlook 2019."*

[35] Hernandez C, O. Mauricio, Shadman, Milad, Amiri, Mojtaba Maali, Silva, Corbiniano, F. Estefen, Segen, La Rovere, Emilio. 2021. "Environmental impacts of offshore wind installation, operation and maintenance, and decommissioning activities: A case study of Brazil." *Renewable and Sustainable Energy Reviews*. 144. https://doi.org/10.1016/j.rser.2021.110994.

[36] Dagoumas, A. S., and Koltsaklis, N. E. 2019. "Review of models for integrating renewable energy in the generation expansion planning," *Applied Energy*. 242:1573-1587, https://doi.org/10.1016/J.APENERGY.2019.03.194.

[37] Lund, Henrik, Thellufsen, Jakob Zinck, Østergaard, Poul Alberg, Sorknæs, Peter, Skov, Iva Ridjan, Vad Mathiesen, Brian. 2021. "EnergyPLAN - Advanced analysis of smart energy systems, Smart Energy, Volume 1." https://doi.org/10.1016/j.segy.2021.100007.

[38] Liu, W., Lund, H., and Mathiesen, B. V. 2011. "Large-scale integration of wind power into the existing Chinese energy system." *Energy* 36:4753-4760, https://doi.org/10.1016/j.energy.2011.05.007.

[39] Cabrera, Pedro, Lund, Henrik, Zinck Thellufsen, Jakob, and Sorknæs, Peter. "The MATLAB Toolbox for EnergyPLAN: A tool to extend energy planning studies." *Science of Computer Programming.*

[40] C. Yongmei, C. Dexin, L. Jun., L. Yanpin, R. Linmao, R. Yan, Z. Hanmin, and Z. Lanjin. *"Distributed wind-light pumped-storage supplementary electricity generation control system and method."* Patent 103490437, January 1, 2014.

[41] C. Haibao, *"System for lifting water through solar photovoltaic and making redundant electricity get on grid,"* Patent 105971836, September 28, 2016.

[42] Chen Peng, Fang Tongqing, Gao Jiang, Li Delu, Liu Hongxia, Liu Hui, Sun Jin, Wang Guoping. *"Seawater desalination device utilizing wind energy and working method thereof."* Patent 106365230, February 1, 2017.

[43] Allen, Mark, Aronson, Neal, Garcia Guerrero, Jose Alberto, Leal Jeldrez, Joan, Walker, Sandra. AU2018251592: *"An integrated system for generating, storing and dispensing clean energy and desalinating water."* Patent AU2018251592, February 1, 2017. October 31, 2019.

[44] Borge Diez, David, García-Moya, Francisco José, Rosales Asensio, Enrique. 2021. "Comprehensive assessment of Gran Canaria water-energy-food nexus with GIS-based tool." *Journal of Cleaner Production* 323. https://doi.org/10.1016/j.jclepro.2021.129197.

[45] Vattenfall. 2021. *"Vattenfall Wind Power Certified Environmental Product Declaration EPD of Electricity fromVattenfalls WindFarms."* Accessed August 2021. http://www.vattenfall.com/en/file/Certified_Environmental_Product_Declaration_of_Electricty_from_Vattenfall.

[46] Serrano-Tovar, Tarik, Peñate Suárez, Baltasar, Musicki, Ana, de la Fuente Bencomo, Juan A., Cabello, Violeta, Giampietro, Mario. 2019. "Structuring an integrated water-energy-food nexus assessment of a local wind energy desalination system for irrigation." *Science of the Total Environment* 689:945-957. https://doi.org/10.1016/j.scitotenv.2019.06.422.

[47] Châtel, F. de., Holst-Warhaft, G., and Steenhuis, T. 2014. *"Water Scarcity, Security and Democracy: A Mediterranean Mosaic."* Global Water Partnership Mediterranean, Ithaca (N. Y.): Cornell University-Atkinson Center for a Sustainable Future.

[48] Keairns, D. L., Darton, R. C., and Irabien, A. 2016. "The energy-water-food nexus." *Annual review of chemical and biomolecular engineering.* 7:239-262.

Chapter 7

Implementation Analysis of the Largest Wind Project in Ecuador

Santiago Pulla Galindo[1,*], David Borge-Diez[2] and Daniel Icaza[1]

[1]Doctoral School, University of León, León, Spain. Catholic University of Cuenca, Ecuador Center for Research, Innovation, and Technology Transfer CIITT, Lighting Technology Research Laboratory CIITT, GIRVyP Group
[2]Electrical, Systems and Automation Engineering, University of Leon, Spain

Abstract

Through the Ministry of Energy and Non-Renewable Natural Resources and the generation Company ELECAUSTRO, the Ecuadorian government, is implementing the "HUASCACHCA" wind power project, considered the largest in the country, with an installed capacity of 50 MW. This project is located in the province of Loja, in the south of the country, in a desert area on the banks of the Jubones river, where the climatic conditions allow maintaining an almost constant airflow throughout the year which will allow the operation of the power plant. This chapter will present the Ecuadorian legal framework that enables the implementation of this type of project and its conditions; it will describe the plant, its operating conditions, and its performance. Since its capacity and state-of-the-art technology will make it a national icon for the application of renewable energies in Ecuador, it can serve as a reference for other projects around the world that are installed in similar conditions.

---

[*]Corresponding Author's Email: gpulla00@estudiantes.unileon.es.

In: Wind Turbines
Editors: David Borge-Diez and Daniel Icaza
ISBN: 978-1-68507-974-1
© 2022 Nova Science Publishers, Inc.

Keywords: wind energy, wind jubones, wind project, wind azuay

Nomenclature

ARCONEL	Agencia de Regulacion y Control de Electricidad (Electricity Control and regulation agency)
BCE	Banco Central del Ecuador (Central Bank of Ecuador)
CENACE	Operador Nacional de Electricidad (National Center for Energy Control)
ELECAUSTRO S.A.	The Generating Company of the Austro
INAMHI	National Institute of Meteorology and Hydrology
INEC	Instituto Ecuatoriano de Estadísticas y Censos (Institute of Statistics and Census of Ecuador)
INEN	Instituto Nacional de Normalización (Ecuadorian Standardization Institute)
kWh	Kilowatt Hour
LOEE	Ley Orgánica de Eficiencia Energética (Law of Energy Efficiency)
LOSPEE	Ley Orgánica del Servicio Público de Energía Eléctrica (Law Electric Power Service)
MAE	Ministerio de Ambiente y Agua (Ministry of Environment and Water)
MEER	Ministerio de Electricidad y Energía Renovable (Ministry of Electricity and Renewable Resources)
MERNNR	Ministerio de Energía y Recursos Naturales no Renovables (Ministry of Energy and Non-Renewable Natural Resources)
MTOP	Ministerio de Transporte y Obras Públicas (Ministry of Transport and Public Works)
MW	Megawatts
MWh	Megawatts-Hour
PME	Plan Maestro de Electricidad 2019 (Electricity Master Plan 2019)
USD	US Dollar
WB	World Bank

Introduction

Renewable energy sources in the course of humanity's existence have been discovered and harnessed in different periods to a greater or lesser extent depending on local situations and existing technological advances. Despite having natural resources, many of them have not been exploited due to local ignorance, lack of technology, or lack of need for usage in the different periods of existence of humanity. Today, society needs to take advantage of these resources from renewable natural ones, even more because energy sources, together with non-renewable resources —mainly those derived from oil—, have already been exploited, and it is a natural resource that is being depleted and because of the climatic consequences that derive from the use of these resources. Currently, there is the technology that, not at a very high cost, allows the use of several of these sources, such as wind, tidal, solar, hydraulic, geothermal, etc. Nowadays, in addition to technological advances in various latitudes of the world, there is a survey of natural resources, which geographically locates the places where certain natural sources can be used, lacking in many cases the political decisions for their implementation.

One of the main obstacles encountered in the implementation of these systems is the social ones, which, either due to the lack of information through adequate dissemination campaigns, the advantages and disadvantages of the projects are not disclosed, finally encountering obstacles to implementation. To these inconveniences, those caused by environmental aspects must also be added, which, in little, medium, or large measure, are presented in the constructions of generation plants. Also, the environmental management of the expected remediation in this case, has not been well disseminated, causing delays or difficulties in the implementation of the project.

Ecuador has not been oblivious to the problems mentioned, both in large-scale projects of mainly hydraulic power plants and to a lesser extent of other types of plants that take advantage of renewable sources.

Despite the difficulties exposed around the world, this type of plant continues to be built with renewable sources to leave the dependence on generation sources that use fossil sources and contribute to the reduction of the greenhouse gas effects that occur on the planet.

There are also legal instruments that, although in some cases are not mandatory for some countries, directly or indirectly promote the generation of electricity using renewable sources. Some instruments were generated in the Kyoto Protocol, adopted on December 11, 1997, in the Paris agreement and on December 12, 2015 and in the different sessions of the "Conference of the

Parties," COP held since 1994. They have established strategies for reducing the emission of greenhouse gases and the planet's temperature, directly and indirectly promoting the use of plants that use renewable generation.

Among the generation plants that use renewable sources are those that use wind, becoming wind power plants. They are installed in sectors that maintain sustainable airflows that allow an almost constant operation without guaranteeing that their process continues to be the one of the disadvantages that these systems present. They also generate effects that are considered environmental problems, such as the presence of noise, changes in the flight paths of birds or insects, and visual impacts, in spite of being frequently used plants around the world, built mainly in isolated places or far from electric power transmission systems.

Ecuador has not been oblivious to the need to adopt this new technology within its plans to implement electricity generation infrastructure. However, its incursion has been to a lesser degree compared to the use of other technologies such as hydraulics or thermal. Although there has been legislation that at the time favored the incursion into these technologies. In recent years, there were even three important projects have been implemented. The most recent one: the Huascachaca Wind Project, is under construction.

This chapter deals with the state of exploitation of wind resources in Ecuador. It recounts the availability and possibilities of using the resource, the legal framework, the current installed capacity, and the projects in execution. Finally, it focuses on the Huascachaca project, which is already addressed in technical details of its wind system, generation, regulation, its installed capacity, its growth prospects, and its electricity sub-transmission system for integration into the interconnected national system of Ecuador. It also addresses the economic aspects of the system and its importance within the national system, emphasizing on the context of the generation systems that Ecuador has. Renewable energies through electricity generation systems have been very little exploited or used in Ecuador due to the importance of addressing the issue of wind energy in an objective way that allows having complete concentrated information in the future. This information can be used to establish policies that would enable making sound decisions in this area. This chapter aims to publicize the state of the situation of wind systems in Ecuador, particularizing the case of the Husacachaca Power Plant and its influence on the economic aspects of the region where it is installed in Ecuador, as well as its geopolitical repercussions. In this chapter, after the introductory part, there is a segment where the leading socio-economic indicators of Ecuador are indicated with updated information to the year 2020, referring mainly to the World Bank, WB [1], Central Bank of Ecuador, BCE [2], Ecuadorian Institute of Statistics and census, INEC [3], etc. Then, the objectives of the chapter are established, indicating that they refer to the presentation of concise and complete information on wind generation systems in Ecuador, framed within a global context that considers the legal, technical, and economic aspects, focused on the Huascachaca wind project for being the most extensive project under construction in Ecuador and the one that will generate more electricity using this type of primary renewable source. After that, the methodology used for the collection of data is described, together with its analysis, using mainly government public information from the Ministry of Electricity and Renewable Energy MEER, Ministry of Energy and Non-Renewable Natural Resources, MERNNR [4], Agency for Regulation and Control of Electricity, ARCONEL, [5], Agency for Regulation and Control of Energy and Non-Renewable Natural Resources, ARCERNNR, [6], National Electricity Operator, CENACE, [7], Generating Company of the Austro ELECAUSTRO, [8], Organic Law of the Public Electric Energy Service[9], and its Regulations, [10], Regulations of the ARCONEL and ARCERNNR, OLADE, [11], World Bank, [1], Central Bank of Ecuador [2], Provincial Government of Azuay, [12], Ministry of Environment, Water and Ecological Transition, MAE, [13], National Institute of Meteorology and Hydrology INAMHI [14], Organic Law of Energy Efficiency, LOEE [15], Ministry of Transport and Public Works, MTOP [16], Latin American Energy Organization, Electricity Master Plan 2019, PME [17], and other references. Later, the results are established in another section, where, with the data collected, the information is analyzed, organized, and exposed in an orderly and concise manner detailing the essential technical aspects of power, number of generators, average annual wind speed, climatic aspects, transmission line, substations, etc., as well as the socio-economic aspects, which arise from the start-up operation, the system's functioning. Then in another section refer to the conclusions that address the main issues analyzed, emphasizing the repercussions that the construction of the Huascachapa project will have at the national level. The possibilities of expansion of these generation systems taking as reference point the legal aspects that regulate these projects, also highlighting the technical, economic, political, environmental, and national integration aspects.

Socio-Economic Aspects of Ecuador

The Republic of Ecuador is located in the South-West of the American continent, with a border to the north and northeast with the Republic of Colombia, to the south and southeast with the Republic of Peru, and to the west with the Pacific ocean. The Galapagos archipelago is in its insular region.

Its territorial extension is 274, 574 km^2, including the Galapagos Archipelago. The country is divided territorially into "Coast" with access to the Pacific ocean, "Sierra," which includes the mountains, volcanic continental part, and is part of the Inter-Andean alley, "Oriente," that is located in the East of Ecuador and is made up of forested and jungle areas. Finally, the region "Insular" or "Galápagos" area consists of islands and islets of the Galapagos archipelago.

The geography is varied, composed of a mountainous area, desert areas —mainly located on the Ecuadorian coast—, and jungle areas. It has important water sources such as Los Ríos, Guayas, Daule, Santiago, Napo, etc. Its main agricultural products are bananas, cocoa, rice, etc. Its necessary mining resources are gold, copper, and mainly oil. Its energy resources are varied, but primarily they are provided by water sources [18].

The population for the year 2021 is estimated at 17,510,643 according to the national census of the year 2010 [3].

The political capital is the city of Quito, and due to its socio-economic aspects, the cities of Guayaquil and Cuenca are also important.

The following are the main macroeconomic references of Ecuador. In Table 1, the information is shown.

Table 1. Economic indicators of Ecuador

Indicator	Value
GPD (year 2020)	539,000 billion USD. [1].
GNI per capita (2020)	5600.39 USD. [1].
Population growth	1.69%.
Unemployment rate	4.4% (Dec. 2021). [3].
Annual inflation index (tear 2020)	0.07% (Dec. 2021). [3].
Consumption per capita annual	(kWh/person) 1147.62 (year 2020). [19].
Electicity cover	97.09% (2020, population.) [19].

As for the aspects of the electricity system in Ecuador, in 2020, a total capacity power of 8,712.29 MW has been installed, with a coverage of 97.09%, with 5,368,697 customers [19].

The main energy aspects related to the electricity service in Ecuador are shown below in Table 2.

Table 2. Energy indicators of Ecuador [19]

Indicator	Value
Gross electricity generation	31,498,80 GWh
Power demand	27,317,09 kWh
Installed capacity	8,712.29 MW
Number of Clients	5,368,697
Marketed energy	20,095.49 GWh
Coverage of the electrical service	97.09%

Objectives

To present the issue of the Huscachaca wind power plant in an orderly and chronological manner and within a global context of energy systems in Ecuador, its study has been proposed with three main objectives:

1. *Present the legal framework, which is the basis for implementing wind generation systems in Ecuador.*
   In Ecuador, in the last ten years, several legal and regulatory instruments were issues that have regulated the implementation of wind generation systems and their operation, maintenance, and cargo dispatch regimes, without neglecting the aspects of energy sales and their costs.

This chapter will present the legal instruments referred to as the Organic Law of the Public Electric Energy Service, [9], and its General Regulations, [10], as well as the relevant state regulations, legal instruments with which the operation and operation of wind generation systems have been regulated.

2. *Detail the technological aspects of the Huascachaca wind power plant*
   All the technological aspects of the Huscachaca wind power plant will be exposed, indicating first the location and the prevailing climatic conditions in the sector. The type of plant to be installed and its components, such as the generation, regulation, control system, and

annexed systems (the lifting substation, transmission line, and the interconnection with the interconnected national system of Ecuador).

3. *Present and analyze the economic aspects of the Huascachaca power plant*
   The Construction Project of the Huscachaca wind power plant bases its implementation not only on the technical aspects but also on the economic ones. Its analysis will be presented in this chapter.

Methodology

Bibliographic research of scientific information has been used to achieve the objectives set out in this chapter. Information has been collected from reference articles, reports, catalogs, laws, regulations, and public information on meteorological aspects. The information deals with the construction details of the project and its aspects and processes of public procurement for the implementation of the system of both its civil works and its electrical work.

The legal information outlined in the chapter has been extracted mainly from the Organic Law of the Public Electric Energy Service [9], its General Regulations [10], and from different state regulations, issued by the control and regulation entities of Ecuador such as the Electricity Regulation and Control Agency, ARCONEL [5], and the Agency for regulation and control of Energy and Non-Renewable Natural Resources, ARCERNNR [6], and regarding the information about the generation systems with non-renewable sources were issued by La Empresa Generadora del Austro ELECAUSTRO S.A. [8], which is a public electricity generation company located in the city of Cuenca and responsible for the design, construction, and project's functioning.

Currently, the system is in the construction process, so information is available from the project's construction contracting in the "SOCE-Official System of Public Procurement of Ecuador" [20], which is a public legal system with internet access, built and based on the mandates of the "Organic Law of the National System of Public Procurement," [21] of Ecuador.

In this paper, one can find the technical information necessary for the construction of the project, both its civil works and the electrical work and its components, as well as the contractual aspects, such as details of the electrical components, the civil works, construction deadlines, costs, etc.

The implementation of his type of project, due to its size and importance within the energy sector of Ecuador, leads to direct and indirect effects on the existing communities within its area of influence. Official reportings to the national and local press of different authorities and local actors regarding the implementation of the project have also been reviewed and included, which shows, to some extent, the related aspects of social communication, which cannot be ignored by the characteristics of influence of these projects.

Results and Analysis

Reference Laws for the Projects

Organic Law of the Public Service of Electric Energy

On January 16, 2015, it was published in the Official Register, Third Supplement No. 418, in the "Organic Law of the Public Service of Electric Energy" [9]. It is a legal instrument that governs, regulates, and controls the activities related to the generation, transmission, distribution, and commercialization of the public service of electric energy and general public lighting in Ecuador from the date of its publication.

This Law was approved by the National Assembly of Ecuador within the framework of the new constitution of the Republic of Ecuador and its constituent mandates, which among others, refer to the following considerations:

That Constituent Mandate No. 9 aims to adopt policies to encourage investment in companies in the electricity sector, which cover the needs of new electricity infrastructure, enabling the use of state resources [9].

That in Constituent Mandate No. 15, significant changes were made in the Ecuadorian electricity sector that achieved, among others, the determination of a single rate, the recognition, through state entities, of the deficit between the costs of the electricity service and those of the rate, the elimination of marginal costs in the valuation of electricity generation costs. Additionally, legislation was passed on the state's contribution in the investment for the expansion of electricity distribution and transmission and the cancellation of commitments of the electricity companies that had been maintained by different transactions of the electric service. Its compliance authorized the development of more projects for the electricity service in the country, the sustainability, and the function of the electricity sector in the country [9].

The article 14 of the Constitution of the Republic of Ecuador emphasizes the rights of the Ecuadorian population to develop in a clean ecologically balanced environment, guaranteeing sustainability and good living for the inhabitants of Ecuador. In addition, the preservation of the environment, the maintenance of ecosystems, the maintenance of biodiversity, the importance of preventing environmental damage, and the need to recover affected natural spaces are declared [9].

Article 15 of the Constitution of the Republic of Ecuador indicates that "it is the responsibility of the State to promote, in the public and private sectors, the use of technologies that are considered environmentally clean and the use of energies that are not polluting and with low environmental impact" [9].

Article 315 of the Republic of Ecuador Constitution states that the ecuadorian state may form public enterprises for the management of strategic areas, for the provision of public services, and the sustainable exploitation of Ecuador's natural resources or public goods [9].

Article 395 of the Constitution of the Republic of Ecuador refers, among others, to the fact that the ecuadorian state must guarantee a sustainable model for development, environmentally balanced, that conserves biodiversity and the natural regeneration of the affected ecological systems [9].

Article 413 of the Constitution of the Republic of Ecuador, indicates, among others, that the ecuadorian state must promote energy efficiency, the implementation and use of clean environmental techniques and technologies, as well as low-impact renewable energies [9].

That, it is imperative to develop a matrix for electricity generation in an economically and environmentally balanced way, promoting the inclusion of clean and renewable energies such as biomass, photovoltaic, tidal, wind, biogas, photovoltaic, geothermal, and hydroelectric, and reducing generation through thermal power plants that are inefficient and whose operation is with the use of fossil fuels [9].

That, it is necessary to have a new legal framework for the ecuadorian electricity sector, which observes the provisions of the new Constitution of the Republic of Ecuador, according to the national reality, updating the institutional conformation of the state [9].

Therefore, and for other considerations, the Organic Law of the Public Service of Electric Energy was approved. It replaced the "Law of Regime of the Electricity Sector," published in the Supplement to the Official Registry No. 43 of October 10, 1996 [22], which had been governing all the activities of the electric energy service in Ecuador.

Among the main aspects established by the new Law and related to the characteristics of electricity generation, environment, and promotion of generation systems through renewable sources, the following are set among the main ones:

The New Law establishes as Object and Scope the following:

Article 1. "The purpose of this law is to ensure that the public electric power service complies with the constitutional principles of obligation, generality, uniformity, responsibility, universality, accessibility, regularity, continuity, quality, environmental sustainability, precaution, prevention, and efficiency, for which, through this instrument, it is appropriate to regulate the exercise of the responsibility of the state to plan, execute, regulate, control and administer the public electricity service. This law regulates the participation of the public and private sectors in activities related to the public electricity service, as well as the promotion and execution of plans and projects with renewable energy sources, and the establishment of energy efficiency mechanisms" [9].

This also proposes nine specific objectives, being those related to generation systems the following:

- "Comply with the provision of the public electricity service to the consumer or end-user, through the activities of generation, transmission, distribution and marketing, and import and export of electrical energy."
- "Develop promotion mechanisms by the state that encourage the technical and economical use of energy resources, emphasizing renewable sources. The promotion of biomass shall have pre-eminence in the origin of solid waste."
- "Formulate energy efficiency policies to be complied with by natural and legal persons who use energy or provide related goods and services, favoring the protection of the environment."
- "Design mechanisms to ensure the economic and financial sustainability of the electricity sector"."
- "Developing rural energy" [9].

In the Law in Title II of "Responsibility is and Attributions of the State," in Article 7 of "Duty of the State" refers to the fact that it is the competence of the central government to make decisions for the planning, construction, and installation of electrical systems to dispose of electrical energy and deliver to final customers, being also responsible for the maintenance of the systems, the operation and sustainable implementation of the ecuadorian electricity sector, to cover the requirements of the public electricity service [9].

Title III of "Structure of the Electricity Sector," in its chapter I, of Structure, in Article 9, it refers to the Institutional Structure of the Electricity Sector and indicates that it is structured by the Ministry of Electricity and Renewable Energy MEER, [4], currently the Ministry of Energy and Non-Renewable Natural Resources MERNNR, by the Electricity Regulation and Control Agency ARCONEL, now the Agency for regulation and control of Energy and Non-Renewable Natural Resources, ARCERNNR, the National Electricity Operator, CENACE, [7], and by other related institutes.

Also Article 10 establishes the Business Structure of the Ecuadorian electricity sector. It indicates that it will act through public companies, private companies, mixed economy, consortia or associations, and popular and solidarity economy companies.

This structure allowed the management of the design, financing, contracting, and construction of the Huascachaca wind project in an orderly technical and environmentally sustainable manner, through the public Company Electro Generadora del Astro.

Article 13, Title III, on the Planning of the Electricity Sector, establishes that the Ministry of Electricity and Renewable Energy is responsible for planning the electricity sector, including renewable energies in a framework of energy efficiency, consistent with the Constitution of the Republic and with the National development plan and national policies issued by the presidency of the Republic of Ecuador. It considers two legal instruments and mandatory compliance in the public sector and referential indicative for the private sector. These two instruments are the Electricity Master Plan, PME [17], and the National Energy Efficiency Plan, PLANNEE [23].

The Electrizad Master Plan, PME, is prepared by the Ministry of Energy and Non-Renewable Natural Resources MERNNR in coordination with companies in the electricity sector of Ecuador.

In this plan, the Huascachaca project has been included since the ten-year Plan from 2017 to 2025, a period in which it has been managed.

The National Energy Efficiency Plan PLANNEE is also prepared by the Ministry of Energy and Non-Renewable Natural Resources in coordination with institutions and secretaries of State related to energy use.

In Title IV of "Management of Non-Conventional Renewable Energy and Energy Sources," in Article 26, of "Non-conventional renewable energies," it indicates that MERNNR must promote the use of technologies considering clean and alternative energies, by what is displayed in the new Constitution that refers to the development of sustainable electricity systems, based on the use of resources considered renewable energy. It also establishes that the electrical energy produced with this type of energy source will have preferential treatment in the regulations issued in the ARCENRRN. This legal instrument allows wind projects, including Huascachaca, to have preferential treatment for the recognition by the state of their production costs per kWh.

Title V of "Provision of the Public Electric Energy Service," in its chapter I of Enabling Titles, in Article, 30, on the "Authorization for the use of renewable and non-conventional energy resources," establishes that the promotion of the use of non-conventional renewable energies, as well as the use of energy resources considered as renewable, must have in advance the authorizations for the use of these resources by, in this case, the National Environmental Authority, and must maintain compliance with the requirements of the governing body in terms of national planning.

Article 40, referring to the "Generation," indicates that this activity of electricity generation can be carried out by different companies: mixed economy companies, public companies, and also by private legal persons, as well as popular and solidarity economy companies, which are correctly trained by the granting authorities for the exercise of this activity. The operations of these companies will be or are under the provisions of their enabling titles, and the constitutional, regulatory, legal, and regulatory instruments established, with their sole responsibility. They must also show principles of transparency, continuity, energy efficiency, and quality. As for the development of new electricity generation projects, these will be based on sectoral policies established in the programs and plans of the sectors and framed in the National Development Plan.

On electricity transactions, Article 46 of "Energy Block Transactions" states that these can be expressly concluded for sales and purchases of energy through contracts signed by the participants. It indicates that they must be settled with the participation of the National Electricity Operator, CENACE [7], based on the prices referred to in the contracts established for these purposes. It is indicated that for the commercial closing of these transactions, settlements of transactions can be made in a short term. The scope and conditions to be referred to in contracts for the sale and purchase of electricity, and short-term trades, are carried out based on the regulations issued for the ARCERNNR [6].

Regarding the programming and operation of the different generation plants, Article 47, "Of the programming of the operation," refers to the fact that the National Electricity Operator, CENACE, must carry out the programming of the operation of the plants in the long, medium and short term, with the prospect of obtaining the lowest operating cost for Ecuador, bearing in mind the technical restrictions.

It refers to the fact that the different participants that carry out actions of generation, self-generation, distribution, commercialization, and transmission, and of the large consumers, and those authorized to carry out different international transactions of sale of electrical energy, must send to the National Electricity Operator, CENACE, the economic reference information, technical data, and toperational plans that must be observed for the programming.

Article 48 of "Of the economic dispatch." indicates that the National Electricity Operator, CENACE, must carry out the economic dispatch of all power generation plants, considered as the basis of the programming of the operation indicated in article 47, to obtain the lowest value of the possible hourly cost of electricity. Variable production costs must be considered, which must be referred to by each electricity generator and audited by such operator, by the respective regulations.

As for the commercial settlements of purchases and sales of energy, these are legislated in Article 49, referring to "he commercial liquidation," and indicates that the sales and purchases of electrical power that are executed between the concessionaires of the electricity sector through reference contracts, in addition to short-term transactions, must be liquidated by CENACE, within its competences, observing the regulations issued by the ARCERNNR. The National Electricity Operator, CENACE, determines the values delivered to each participant.

Regarding short-term transactions, Article 51 indicates that a "short-term transaction" should be considered, which may be presented by the difference between the amounts of energy consumed or produced, and the amounts actually contracted. It indicates that the electricalpower must be valued at an economical cost determined by the actual dispatch of electricity generation at the end of each hour, known with the name of the hourly cost of electrical energy.

Regarding investment and planning for the Ecuadorian electricity sector, Article 53.- indicates that the PMD Electricity Master Plan [17], which is currently carried out by the Ministry of Energy and Non-Renewable Natural Resources, is carried out for a ten-year projection, identifies each electricity generation project to be developed. This Plan also identifies the expansion and improvement projects of the electricity generation systems.

This article also indicates that the current Ministry of Energy and Non-Renewable Natural Resources will select from the Electricity mentioned above Master Plan [17]. The projects to be developed by the Ecuadorian State and others that may be proposed to other private companies or popular and solidarity economy, for which there must be a public process referred to in the Law.

The investments required to develop SME generation projects by other entities and public companies can be made from the general state budget or through these companies' resources. It is also indicated that, for the financing of projects, companies can contract or obtain credits with their guarantees or given by the Ecuadorian State.

As for the rate regimes, these are set out in chapter III, in eight articles, highlighting the following, and that refer to generation by renewable sources.

Article 54 of "Prices subject to regulation rates" indicates that currently, the ARCERNNR, in the first half of the year, must determine the costs of the generation that must be applied in the electrical transactions. These will serve, among others, as a basis to determine the electric rates that must be applied in the following year to the consumer or end-user.

This article also indicates that the current ARCERNNR, in the face of a previous study, can set rates whose purpose is to promote and encourage the implementation of primary industries, bearing in mind the use of environmentally friendly renewable energies, also considering that it must be at competitive prices, or that they can be subsidized if necessary.

Regarding the cost of the public electric energy service, in Article 56., and referring to the costs of generation, it is indicated that it corresponds to the value that the consumer or user has to pay for the supply of electrical energy, the same that must cover the costs that arise from the generation activity that must be operated optimally.

Regarding the companies of generation of popular and solidarity economy or private, the costs must consider the assets immersed in the service and the costs for the administration, maintenance, and operation; and costs for environmental responsibility.

In the case of public and mixed generation companies, the costs must consider the items for aspects of reliability, quality, availability, administration, maintenance, and operation, and the costs referred to as environmental responsibility.

In the case of electricity generators of public companies, it is indicated that 30% of the surplus released in the operation phase will be destined for other projects in the territorial development of the area of influence of the project in operation. In the case of mixed or private economy capital generators, it is indicated that 3% of the profits will be allocated to workers from the validity of the law. The remaining 12% will be given to other territorial development projects in the project's area of influence.

Regarding Energy Efficiency, Title VI, on Energy Efficiency, in Article 75 of "Establishment of energy efficiency policies," it refers to the fact that the policies and standards applied by the Ministry of Energy and Non-Renewable Natural Resources to meet the objectives of this matter, must promote greater energy efficiency in the exploitation of energy sources and the use of electrical energy.

Title VII of the Law refers to "Environmental Responsibility" and is made up of four articles, of which we highlight the following:

Article 77 of "Coordination" refers to the competence in the coordination of the "National Environmental Authority," which will be responsible for monitoring compliance with the regulations that must be complied with by the electricity companies.

Article78, of "Protection of the environment" indicates that it is the responsibility of electricity companies, whether private or of popular and solidarity economy, mixed, public, and all participants in the electricity sector that carry out generation activities, and others, to comply with all applicable policies, standards and procedures established by the National Environmental Authority, in terms of prevention, mitigation, control, repair and monitoring of environmental impacts in the different stages of construction, operation and then retirement.

Article 80 of "Environmental impacts," indicates that electricity companies must prevent, mitigate, and/or remedy or compensate, as the case may be, the impacts considered harmful that may occur in the environment by the activities of construction, operation, and maintenance of electrical generation systems and others.

The Law regarding the Declarations of utility and easement of transit necessary for the development of activities concerning the generation plants indicates Title VIII, of "Declarations of public utility and transit easements," in the related terms indicates:

Article 81 of "Declarations of public utility" indicates that the Ministry of Energy and Non-Renewable Natural Resources or public companies that provide the public service of electric energy, for express reasons of public utility or ecuadorian social and national interests, may buy real estate, before the declaration as an asset of public utility or social welfare, framed in the Constitution and the Organic Law of the National System of Public Procurement, which is necessary for the particular execution of generation activities.

The Ministry of Energy and Non-Renewable Natural Resources, at the request of legal persons, whether private or of popular and solidarity economy, may process the declaration of public utility or social and national interest real estate, which is required for the implementation of electricity generation activities.

Finally, in the "General Provisions," of the Law, the second provision of "Binational Generation," indicates that the Ecuadorian State can implement new electricity generation projects in a binational manner and that the administration of this generation must be carried out based on agreements, and binational legal contracts or instruments.

The Law was signed in the National Assembly on January 8, 2015, and is in force at the publication of this chapter.

General Regulation of the Organic Law of the Public Service of Electric Energy

For an adequate application of the Organic Law of the Public Service of Electric Energy [9], on August 20, 2019, Executive Decree 856 was published in the Official Registry. By its means, the "General Regulation of the Organic Law of the Public Service of Electric Energy" is issued [10], the same that defines the particularities and processes for the adequate application of the Law.

The Regulation in the parts related to the generation systems indicates the following relevant aspects:

In TITLE II, on the "STRUCTURE AND FUNCTIONING OF THE ECUADORIAN ELECTRICITY SECTOR," Chapter II, Article 13- of "Planning the expansion of the electricity sector," states that the Ministry of Energy and Non-Renewable Natural Resources, in coordination with the different entities of the Ecuadorian electricity sector, must carry out the planning of generation projects for the expansion of the sector, among others. They must be executed to supply the demand over time, for which concepts of safety, reliability, efficiency, and socio-environmental responsibility must be considered.

To fulfill this purpose, the "Ministry of Energy and Non-Renewable Natural Resources" must keep updated the inventories of energy resources for electricity generation purposes and must manageopportunity and the studies both from pre-feasibility and feasibility of possible generation projects.

Article 15, on the planning processes related to expansion, refers to the fact that the Ministry of Energy and Non-Renewable Natural Resources will be responsible for planning the generation stage, which includes generation with non-conventional renewable energy sources.

In Chapter III, of "Regime of Participation in the Electricity Sector," concerning to generation, the following is indicated:

Article 18, on "Business participation in generation," refers to the fact that electricity generation activities may be carried out by public, private, mixed economy, associations or consortia, and popular and solidarity economy institutions, which must act under the provisions of the LOSPEE, the "Organic Law of Public Companies," and the "Law of Popular and Solidarity Economy," and other laws applicable in the field, and these regulations mustfollow the Enabling Titles approved by the "Ministry of Energy and Non-Renewable Natural Resources"About companies incorporated under the "Companies Law" [24], which include foreign state-owned companies or their subsidiaries and companies of a consortium or mixed economy type —where such companies have a majority stake— it is demonstrated that the provisions of the regulations of the electricity sector regarding private companies are applied, with the exceptions indicated in the LOSPEE. Majority participation is defined as one in which the foreign country has over 50% of the capital.

Article 19, which refers to "The participation of public and mixed companies," indicates that the Ministry of Energy and Non-Renewable Natural Resources may authorize public companies that have been created by the executive function of the ecuadorian state to participate in the electricity generation activity, and that have within their social objectives only the generation of energy.

It alludes to the fact that, if the ecuadorian state cannot execute or implement electricity generation projects with public companies, the "Ministry of Energy and Non-Renewable Natural Resources" may provide for the participation of mixed economy companies —where the Ecuadorian State has a majority stake— becuase they contain the actions of energy generation in their sole social objective.

Also, the "Ministry of Energy and Non-Renewable Natural Resources" may provide participation in electricity generation activities to public or mixed institutions made up of the "Decentralized Autonomous Governments," where they have majority participation, only if the projects that the aforementioned companies may present are observing the policies established by the "Ministry of Energy and Non-Renewable Natural Resources," in planning the growth of generation systems.

Article 20, on "Private participation and the popular and solidarity economy," refers to the fact that the "Ministry of Energy and Non-Renewable

Natural Resources" may delegate private companies or companies of popular and solidarity economy, action in activities of the electricity generation sector, in addition to the participation in the projects or generation blocks identified in the PME, becuase of the conditions in Article 25,, which are met in paragraphs one and two of the LOSPEE Public Processes for selection - PPS will be carried out for the objective . Article 21, on "Participation in generation projects with Non-Conventional Renewable Energy, NCRE, provided for in the PME," indicates that the "Ministry of Energy and Non-Renewable Natural Resources," in response to the "National Development Plan" and the established sectoral policies, will consider the planning, implementation of NCRE projects, and promote its performance through private companies of popular and solidarity economy. The generating companies that are favored will receive for the cost of the energy generated, the price of the energy that has been established in the public selection processes PPS, during the entire period that the concession lasts, and will have conditions for the dispatch of preferential energy according to the provisions of the regulations issued by the ARCERNNR. Article 22.- on "Participation in generation projects with NCRE not foreseen in the PME," refers to the fact that "The Ministry of Energy and Non-Renewable Natural Resources" may grant the implementation of NCRE projects that are not foreseen in the Electrification Master Plan, raised either by private initiative or by companies of popular and solidarity economy, for the following topics:

1. "For the sale of energy at regulated demand," the interested parties must submit the project to the "Ministry of Energy and Non-Renewable Natural Resources," who ""will determine if the referred project can be of public interest.
2. About the sale of electricity to large consumers, it shall be by bilateral contracts, provided that the grantor refers that the identified generation projects are consistent with the planning of the sector.

Current Situation of Renewable Energy Systems in Ecuador

Ecuador currently has an electricity generation park that mainly uses renewable sources. This information can be found in the statistics available to the Ministry of Energy and Non-Renewable Natural Resources through the ARCERNNR, which is issued in an updated manner year after year. At the publication of this chapter, ARCERNNR has updated information for 2020 and indicates that it has a gross generation of 31,248 GWh. [19]. It has a nominal power generation from renewable sources of 5,299.09 MW, and in nominal power of energy with non-renewable sources of 3,413.21 MW, this is 60.82% and 39.18%, respectively [19].

5,098,75 MW of this total corresponds to hydraulic generation, which corresponds to 58.52% of the total power generation. The other renewable generation sources correspond to Wind with 21.15 MW, photovoltaic with 27.63 MW, biomass with 144.30 MW, and biogas with 7.26MW [19]. Table 3 and Figure 1 show the different components of both renewable and non-renewable sources.

Table 3. Sources of Electricity Generation of Ecuador year 2020

Electrical energy	Installed power in generation				Production and imports				Delivered to public service	
					Total		Only SIN			
	Nominal Power		Effective power		GWh	%	GWh	%	GWh	%
	MW	%	MW	%	31,248.80	100.00	27,551.32	100.00	25,855.09	100.00
Renewable non renowable	8,712.29	100.00	8,095.25	100.00	31,248.00	99.20	27,300.52	99.09	25,604.29	99.03
Renewable non renowable	5,299.09	60.82	5,524.95	64.91	24,918.71	79.11	24,888.89	90.34	23,444.65	90.68
Hydraulic	5,098.75	58.62	5,064.16	62.56	24,333.26	77.25	2,431.85	88.25	23,107.39	89.37
Wind	21.15	0.24	21.15	0.26	77.10	0.24	71.64	0.26	75.23	0.29
Photovoltaic	27.63	0.32	26.74	0.33	37.76	0.12	33.82	0.12	37.19	0.14
Biomass	144.3	1.66	136.40	1.68	426.59	1.35	426.59	1.55	181.21	0.70
Biogas	7.26	0.08	6.50	0.08	43.99	0.14	43.99	0.16	43.62	0.17
Non Renewable	3,413.21	39.18	2,840.30	25.09	6,329.29	20.09	2,411.63	8.75	2,159.64	8.35
MCI	2,029.74	23.30	1,633.25	20.18	4,422.11	14.04	783.10	2.84	623.58	2.41
Turnogas	921.85	10.58	775.55	9.58	981.75	3.12	708.16	2.57	686.77	2.66
Turbovapor	461.63	5.30	431.50	5.33	925.43	2.94	920.37	3.34	849.29	3.28

Source ARCERNNR [19].

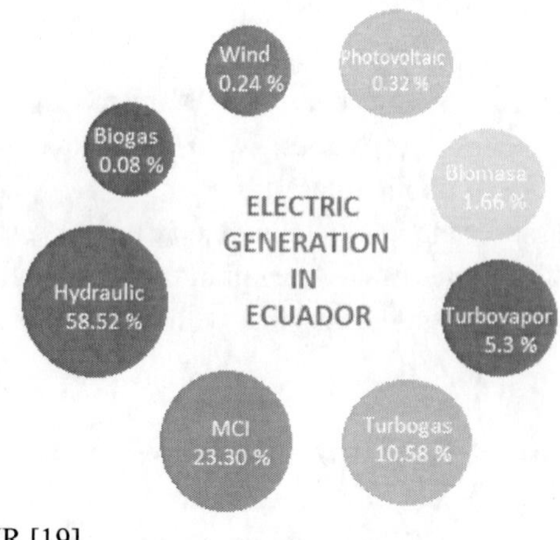

Source ARCERNNR [19].

Figure 1. Components by generation source type.

From the tables above, it is observed and concluded that 0.24% of the power in generation sources corresponds to wind sources. The Ecuadorian State in the Electrification Master Plan 2018-2027 [17], and its scope until 2031 [25], has planned to develop several generation projects, with renewable sources including some wind, being planned the Minas Huascachaca project, with a capacity of 50 MW, and with a generation potential of 132.90 GWh/year. It is additionally scheduled to build the wind generation projects indicated in the following Table 4:

Table 4. Wind Generation Projects in Ecuador until 2031 [25]

Year	Project	Generation	Power [MW]	Energy [GWh/year]
2021	Minas de Huascachaca	Wind	50	132,9
2023	Villanaco III	Wind	110	384
		Total	160	516.9

Minas Huascachaca Wind Project Details and Analysis

The Wind Generation Project called Minas-Huascachaca is built by the Company ELECAUSTRO, which to execute the project and by the requirements of the Organic Law of the Public Service of Electric Energy, and its General Regulations, processed and has the Enabling Title for the "Construction, installation, and operation of the Huascachaca Mines Wind Project of 50 MW of Capacity." It was granted in the resolution of the Agency for Regulation and Control of Electricity N° ARCONEL 026/15, which was notarized, by public deed on February 22, 2016.

- The project is estimated to generate 126-kilowatt hours/year and reduce CO_2 emissions by 76,625 tons. per year [26], if it is built.
- The cost of the project is estimated at $. 101,000,000.00, financed by the Development Bank of Ecuador BDE [27], for $ 54,000,000,000.00 and the rest by own resources of ELECAUSTRO S.A.
- The project is located in the Parish of San Sebastián de Yuluc, in the Saraguro Parish, in the Province of Loja.

In Figure 2 the detail of the project at the entrance to it is shown.

Source ELECAUSTRO [8].

Figure 2. Minas de Huascachaca Wind Project.

The project is located southwest of the equator between the UTM WGS 84 coordinates shown in Table 5.

Table 5. Location UTM WGS 84 [28]

E	N
676,960	9,629,954
682,240	9,629,953
681,120	9,626,763
676,950	9,626,763

Source ELECAUSTRO [8].

Figure 3. PEMH project location.

The project is located between 920 meters above sea level and 1200 meters above sea level, altitude; located between semi-desert ravines. Its temperature is between 20°C and 35°C [28]. The sector has very little vegetation and is considered arid. The project will directly and indirectly benefit the populations of Sumaipamba, Tuluc, Uchucay, and the population of the Santa Isabel peopleish. Figure 3 shows a photo of the project location.

Before the construction, the ELECAUSTRO Company carried out pre-feasibility, feasibility, and other complementary studies, which included analysis of climatological measurements, geological studies, studies for access roads, studies on wind potential, and energy studies, equipment, electrical interconnection, environmental impact studies, and project costs.

The final designs of the transmission line and substations for the Minas de Huascachaca wind project were also carried out. In this context, the UCHUCAY substation was designed. It is is the collector and is located in the wind farm, a transmission line that works at 138 kV and that will connect the wind farm with the National Interconnected System of Ecuador SIN, the design of the "La Paz" substation, which will allow interconnection with the Cuenca-Loja transmission line of the CELEC EP Company, [28]. As for the wind speed and according to the data obtained in the mentioned stations, these would be between 5 and 6 m/s.

It also highlights the final Environmental Impact Study of the Wind Farm, which culminated in obtaining Environmental License No. 078/13 for the stages of construction, operation, and removal of the PEMH. As can be seen, the Minas Huascachaca project is the largest wind generation project currently under construction and is expected to be put into operation by 2022.

In summary, the integral project consists of the construction of 2 Uchucay and La Paz substations, Ushucay La Paz Transmission line at 138kV in an approximate length of 22.5 km, the supply transport and assembly of 15 wind turbines of 3.3 MW, foundries for the wind turbines with their support towers, platforms around the bases for the wind turbines, access roads, collector circuit, construction of platforms and foundations for wind turbines, etc. Figure 4 shows a photo of the Ushucay substation.

In addition, the ELECAUSTRO Company installed four measuring stations that provided the necessary information for the pre-feasibility studies; these are the Ushucay 40, Uchucay 50, Yuluc 50, and Yuluc 80 stations [29].

Within the project, it is also planned to build road works to access the plants, substations, wind turbines, and the construction of two bridges over the Jubones River, which are necessary for implementing the project.

Source MERNNR [26].

Figure 4. Uchucay substation, belonging to the Minas de Huascachaca wind project.

For the implementation of the project, the State Company, Electro Generadora del Austro ELECAUSTRO, owner of the project, proceeded to make several contracts for the provision of equipment, construction of civil works, and necessary transport. The outstanding ones, among others, were made by:

Contract No. 2020-021, dated April 29, 2020, "MANUFACTURE, SUPPLY, TRANSPORT, AND DELIVERY IN ECUADORIAN PORT OF WIND TURBINES, TOWERS AND ASSOCIATED EQUIPMENT FOR THE HUASCACHACA MINES WIND FARM OF 50 MW," this contract, which by its amount, is the largest of all the others was signed with the Company "Dongfang Electric International Corporation," with residence in the People's Republic of China.

This contracthas been developed for the provision of components, parts, and equipment related to the supply of the 15 wind turbines, supply of foundation cage, manufacture of the towers for the installation of the wind turbines, execution of tests and tests in the factory, transport of factory to ecuadorian port, supply of software of the equipment supplied, delivery of assembly manuals, supply of spare parts, among others.

The Contract was made for $ 45,680,000.00, and its process code in the official public procurement system [30] of Ecuador is the SIB-EEGA-001-2020, in which it is registered.

Before the award of the contract, the ELECAUSTRO Company carried out an analysis of the offers of economic feasibility and return on investment of the "Evaluation report of the offers presented within the selection procedure abroad for the contracting of Segment 1 of Phase TWO B of the Minas de Huascachaca Wind Project". It was based on the data presented by the bidders, and others, such as wind turbine costs, assembly costs foundation costs, accesses and platforms, annual investments, annual energy losses, energy sales prices, annual energy in MW-h/year, gross income, net income, cash flow, net annual cash flow, and calculating the NPV for each proposal. It was determined that the proposal of the DONG – FANG company was the most favorable with the Net Present Value VAN of $32,358,309.68 [31], demonstrating the economic feasibility and return on investment, so it was recommended to contract with the referred company and that it was finally awarded.

The project is planned to operate for 20 years and is expected to start operations in 2022.

Contract No. 2021-0023, signed on July 20, 2021, "HANDLING IN ECUADORIAN PORT, LAND TRANSPORTATION TO THE PROJECT SITE, ASSEMBLY, INSTALLATION, TESTING, COMMISSIONING AND START OF OPERATION AND OPERATION AND OPERATION AND MAINTENANCE FOR TWO YEARS OF WIND TURBINES, TOWERS AND ASSOCIATED EQUIPMENT OF THE PEMH OF 50 MW"is based on an amount $11,400,000.00., and was signed with the Company "DONGFANG ELECTRIC ECUADOR S.A." [32].

Its objective isto transport material from an ecuadorian port to the place of assembly of the equipment arrived from China. Instruments like wind turbines, assembly of support towers that are 90 meters high, support equipment, assembly, and installation and execution of tests, in addition to the commissioning and operation of the 50 MW wind project, as well as the maintenance and function for a period of two years of the wind turbines.

Conclusion

The conclusions are presented around four points of analysis for a proper approach to them. They are: (1) the existing legal framework in Ecuador that serves as the basis for the implementation of renewable generation systems, including wind power in Ecuador; (2) the technological details of the Huascachaca wind power plant; (3)the economic aspects that justify the construction of the Minas Huascachaca wind power plant; and (4) the importance of the project within the global environment of the generation sector with renewable sources in Ecuador.

The planning of renewable generation systems in Ecuador, their pre-feasibility, feasibility, and definitive designs, financing, concession, contracting, construction, control, operation, maintenance, and projects' operation, are currently in relation with what is defined in the "Organic Law of the Public Electric Energy Service," which was published on January 15, 2015, its General Regulations, and by the respective regulations and resolutions issued by the Agency for Regulation and Control of Energy and Non-Renewable Natural Resources, ARCERNNR, within the framework of the law.

The legal framework refers to the fact that a generation plant, including wind power plants, can be identified, planned, designed, financed, built, operated, and maintained either by public companies, private companies or joint ventures, or by popular and solidarity economy. They must manage an "Enabling Title"— a legal instrument that empowers the different activities aimed at the design, construction, operation, and maintenance of a plant—, the same that is granted before compliance with the requirements of the law by the energy regulator ARCERNNR.

All generation projects must be included within the "Electricity Master Plan," which is a ten-year plan, and issued by the ecuadorian state through the Ministry of Energy and Non-Renewable Natural Resources MERNNR.

The LOSPEE Law also states that the rate specifications for the purchase and sale of energy, and the promotion policies for the implementation of generation plants with renewable sources, will be issued by the Agency for Regulation and Control of Energy and Non-Renewable Natural Resources ARCERNNR.

The LOSPEE law, and its general Regulations, states that the dispatch of the plants, including those for wind generation, will be managed by the "National Electricity Operator" CENASE and that the energy dispatched will be liquidated by it, following the legal provisions for this purpose. CENASE is a state entity created by the legal mandate of LOSPEE. It is a strategic entity which, also operates to manage the commercial and technical operation of Ecuador's "Sistema Nacional Interconectado" SNI (by its Spanish acronym) and the necessary international interconnections.

In the case of the Minas Huascachaca wind power plant, the company ELECTRO GENERADORA DEL AUSTRO ELECAUSTRO S.A. is in charge of the different procedures for its construction. This company is constituted with shareholders with public capital and is a company legally constituted on August 27, 1999. Its Mission is: "ELECAUSTRO is an innovative and sustainable company that contributes to the country's development through the production of renewable and diversified electricity, basing its actions on ethical principles, quality, and responsibility to society and the environment."

The Company ELECAUSTRO is legally based and has its headquarters in the city of Cuenca del Ecuador since its formation. It managed and obtained the enabling title for executing the Minas de Huascachaca wind power plant, which was granted by resolution of the Agency for Regulation and Control of Electricity No. ARCONEL 026/15, which was notarized, by public deed, on February 22, 2016. The project has environmental impact studies, highlighting the definitive environmental impact of the wind farm, which culminated obtaining the Environmental License No. 078/13, for the stages of construction, operation, and removal of the PEMH.

The Minas wind power plant project in Huascachaca is identified within the 2018-2027 Electricity Master Plan, has an enabling title, has feasibility and definitive studies, as well as the respective environmental studies and licenses, and has passed the socialization processes, including the respective financing, etc. The procedures for the plant start-up meet all the requirements of the Law. In addition, it, withing its execution, has been defined to include access roads and the construction of two bridges over the Jubones river adjacent to the plant.

The Minas Huascachaca wind power plant is located in the South West of Ecuador, in the Province of Loja, at an altitude between 920 meters above sea level and 1200 meters above sea level, located between semi-desert ravines. The average temperature is between 20°C and 35°C. The Minas Huscachaca wind power plant which is one of 50 MW capacity, will be able to generate 126 kilowatts / hour / year, and it is estimated to reduce CO_2 emissions by 76,625 tons per year.

The project will, directly and indirectly, benefit the populations of Sumaipamba, Tuluc, Uchucay, and the population of the Santa Isabel parish.

The cost of the project is estimated at $. 101,000,000.00, financed by the Development Bank of Ecuador BDE with $. 54,000,000.00 and the rest with resources of the Company ELECAUSTRO S.A.

The project includes the installation of 15 towers with 3.3 MW wind turbines.The towers are 90 meters high, and the blades of the wind turbines are 72 meters long, becoming the most extensive plant installed in Ecuador. The system includes the construction of two substations called UCHUCAY and La Paz, as well as a 138 kV transmission line, in an approximate length of 22.5 km for the interconnection to the National Interconnected System SIN of Ecuador. The energy generated by each wind turbine is transmitted through a network or collecting system at 34.5 kV to the UCHUCAY substation.

It also highlights the final environmental impact study of the wind farm, which culminated obtaining the Environmental License No. 078/13, for the stages of construction, operation, and PEMH removal. Within the execution of the project, it has also been defined to include access roads and the construction of two bridges over the Jubones river, adjacent to the plant. As shown before, the Minas Huascachaca project is currently the largest wind generation project under construction and is expected to be put into operation by 2022.

An analysis has been carried out in the "Evaluation report of the bids presented within the selection procedure abroad for the contracting of Segment 1 of phase TWO B of the Minas of Huascachaca Wind Project," based on the data presented by the bidders and others. The analysis included costs of wind turbines, assembly costs, foundation costs, access and platforms, annual investments, annual energy losses, energy sales prices, annual energy in MW-h/year, gross revenue, net income, cash flow, and net annual cash flow, where the NPV was determined for each construction bid. It was determined that the proposal of the DONG – FANG Company is the most favorable, with a Net Present Value VAN of $. 32,358,309.68.

Up to date, the Minas de Huascachaca wind project is the largest in Ecuador under construction, with a capacity of 50 MW. Within the Electricity Master Plan in Ecuador, it is plannedthat the wind energy projects of "Villanaco III," with a gross capacity of 110 MW, generating 384 GWh / year, would be implemented by 2031, together with its commissioning. This project, being the only additional to Minas Huascachaca, is planned to be built by 2031. Despite the the two aforementioned projects start-up, the use of wind potential in Ecuador can be considered the marginal compared to other sources of generation in Ecuador. Although the water potential favors the construction of hydraulic power plants in Ecuador, it has not ruled out using other natural potentials, diversifying the use of generation sources, which will finally allow greater reliability of the Ecuadorian electricity system.

Acknowledgments

The authors thank the CIITT center for research, innovation and technology transfer, the Scademic unit of industrial and construction engineering, the Faculty of electrical engineering, of the Catholic University of Cuenca, and the Company ELECTRO GENERADORA DEL AUSTRO ELECAUSTRO S.A.

References

[1] BM, 2020. *The Word Bank*. Available at: https://www.worldbank.org/en/country/ecuador (accessed in August 10 of 2020).

[2] BCE, 2020. *Banco Central del Ecuador* [*Central Bank of Ecuador*] – Available at: https://www.bce.fin.ec/index.php/boletines-de-prensa-archivo/item/1359-la-econom%C3%ADa-ecuatoriana-creci%C3%B3-01-en-2019 (accessed in July 15 of 2020).

[3] INEC, 2020. Instituto nacional de estadística y censos [National Institute of Statistics and Census]. Available at: https://www.ecuadorencifras.gob.ec/estadisticas/ (accessed in July 28 of 2020).

[4] MERNNR, 2020a. Ministerio de Energía y Recursos Naturales No Renovables [Ministry of Energy and Non-Renewable Natural Resources]. Available at: https://www.recursosyenergia.gob.ec/ (accessed in June 15 of 2020).

[5] ARCONEL, 2020a. Agencia de Regulación y Control de la Electricidad [Electricity Regulation and Control Agency], Available At: https://www.regulacionelectrica.gob.ec/ (accessed in July 12 of 2020).

[6] Agencia de Regulación y Control de la Enerrgía y Recursos Naturales No Renovbales [Agency for the Regulation and Control of Energy and Non-Renewable Natural Resources]. Available At: https://www.gob.ec/index.php/arcernnr/ accessed in December 2021).

[7] CENACE, 2020. *Operador Nacional de Eléctricidad* [*National Electricity Operator*]. Available At: http://www.cenace.org.ec/ (accessed in Julio 22 of 2020).

[8] Empresa Generadora del Austro [Austro Generating Company] S.A. Available At: https://www.elecaustro.gob.ec/ proyectos/proyecto-eolico-minas-de-huascachaca/ Accessed in December 2021).

[9] LOSPEE, 2015. *Ley Orgánica del Servicio Público de Energía Eléctrica* [*Organic Law of the Public Service of Electric Power*], año 2015. Available At: https://www.regulacionelectrica.gob.ec/ley-organica-de-servicio-publico-de-energia-electrica-da-paso-a-la-creacion-de-la-arconel/ (accessed in August 12 of 2020).

[10] Ministerio de Energía y Recursos Naturales No Renovables [Ministry of Energy and Non-Renewable Natural Resources], 2019, "Reglamento General de la Ley Organica del Servicio Público de Energía Eléctrica," ["General Regulations of the Organic Law of the Electric Power Public Service,"] Available At: **Error! Hyperlink reference not valid.**+LA+LEY+ORG%C3%81NICA+DEL+SERVICIO+P%C3%9ABLICO+DE+ENERG%C3%8DA+EL%C3%89CTRICA/d3f53b87-ac86-4fca-90fc-93f4fdd6534a (accessed in November of 2021).

[11] OLADE 2020. Organización Latinoamericana de Energía. INFRAESTRUCTURA DE LA CALIDAD PARAPROGRAMAS DEEFICIENCIA ENERGÉTICA en América Latina y el Caribe, Mayo 2019. [Latin American Energy Organization. QUALITY INFRASTRUCTURE FOR ENERGY EFFICIENCY PROGRAMS in Latin America and the Caribbean, May 2019.] Available at: **Error! Hyperlink reference not valid.**opac-tmpl/Documentos/old0425.pdf (accessed in Julio 29 of 2020).

[12] Gobierno Autonomo Descentralizado del Azuay [Decentralized Autonomous Government of Azuay], Available At: https://www.azuay.gob.ec/ (accessed in octubre of 2021).

[13] MAAE, 2020. Ministerio de Ambiente y Agua [Ministry of Environment and Water]. Available at: https://www.ambiente. gob.ec/ (accessed in Julio 20 of 2020).

[14] Instituto Nacional de Metereología e Hidrología [National Instutite of Meteorology and Hydrology] INAMHI. Available At: http://www.inamhi.gob.ec/ accessed in October 2021.

[15] LOEE, 2019. Ley Orgánica de Eficiencia Energética [Organic Law on Energy Efficiency], año 2019. Available At: https://www.recursosyenergia.gob.ec› downloads›2019/03›L. (accessed in different date).

[16] MTOP, 2020. Ministerio de Transporte y Obras Publicas [Ministry of Transport and Public Works]. Available At: https://www.obraspublicas.gob.ec/ (accessed in Julio 25 of 2020).

[17] PMD, 2020. Plan Maestro de Electricidad [Electricity Master Plan] 2019-2027 Available At: https://www. recursosyenergia.gob.ec/plan-maestro-de-electricidad/ (accessed in Agosto 5 of 2020).

[18] Ministerio de Relaciones Exteriores del Ecuador [Ministry of Foreign Affairs of Ecuador] 2020, Available At: http://www.exteriores.gob.es/Documents/FichasPais/ECUADOR_FICHA%20PAIS.pdf (accessed in september 2021).

[19] Agencia de Regulación y Control de la Energía y Recursos Naturales No Renovables [Agency for the Regulation and Control of Energy and Non-Renewable Natural Resources], 2020 "Estadisticas del Sector Eléctrico Ecuatoriano [Statistics of the Ecuadorian Electricity Sector] 2020" Available At: https://www. controlrecursosyenergia.gob.ec/estadistica-del-sector-electrico/ (accessed in september 2021).

[20] *Servicio Nacional de Compras Públicas* [*National Public Procurement Service*]. Available At: https://portal. compraspublicas.gob.ec/sercop/ (accessed in December 2021).

[21] *Ley Organica del Sistema Nacional de Contratación Pública* [Organic Law of the National Public Procurement System]. Available At: **Error! Hyperlink reference not valid.**actualizada1702.pdf (accessed in november 2021).

[22] *Registro Oficial del Ecuador* [*Official Registry of Ecuador*]. Available At: https://www.registroficial.gob.ec/index.php (accessed in november 2021).

[23] Plan Nacional de Eficiencia Energética [National Energy Efficiency Plan], PLANNEE. Available At: https://energia. gob.cl/consultas-publicas/plan-nacional-de-eficiencia-energetica-2022-2026 (accessed in november 2021).

[24] *Ley de Compañias* [*Companies' Law*], At: https://www.gob.ec/regulaciones/ley-companias (accessed in october 2021).

[25] *Alcance del Plan Maestro de Electricidad* [*Scope of the Electricity Master Plan*] Available At: https://www.recursosyenergia.gob.ec/wp-content/uploads/2021/08/Ajustes-tablas-PME.pdf (accessed in december 2021).

[26] Ministerio de Energía y Recursos Naturales No Renovabes [Ministry of Energy and Non-Renewable Natural Resources], 2021, Available At: 80% de avance presenta Minas de Huascachaca, el proyecto eólico más grande del Ecuador [80% progress presents Minas de Huascachaca, the largest wind project in Ecuador] – Ministerio de Energía y Recursos Naturales no Renovables [Ministry of Energy and Non-Renewable Natural Resources] (recursosyenergia.gob.ec) (accessed in november 2021).

[27] ELECAUSTRO, 2021, Available At: https://www.elecaustro.gob.ec/proyectos/proyecto-eolico-minas-de-huascachaca/ (accessed in octover 2021).

[28] ELECAUSTRO Pliego de Contratación "Fabricación, Transporte marítimo y entrega en puerto Ecutoriano de Aerogeneradores, Torres y Equipos asociados para el Parque Eólico Minas de Huascachaca de 50 MW," [Contract Specification "Manufacture, maritime transport and delivery in the Ecuadorian port of Wind Turbines, Towers and associated Equipment for the 50 MW Minas de Huascachaca Wind Farm,"] 2019, Available At: https://www.compraspublicas.gob.ec/ProcesoContratacion/compras/PC/informacionProcesoContratacion2.cpe?idSoliCompra=DID0RfCl_XpGnVtVUqGUD1jfFosOdciHdB_n-LmL_rw (accessed in octuber 2021).

[29] Francisco Xavier Andrade Rojas,Tesis Maestria, 2015, "MODELO MULTIESTADO PARA ESTUDIOS DE CONFIABILIDAD DEL PARQUE EÓLICO MINAS DE HUASCACHACA." ["MULTI-STATE MODEL FOR RELIABILITY STUDIES OF THE MINAS DE HUASCACHACA WIND FARM."]

[30] ELECAUSTRO, *"Contrato Nro* [*Contract No.*] *2021 -0021,"* Available At: https://www.compraspublicas.gob.ec/ProcesoContratacion/compras/PC/informacionProcesoContratacion2.cpe?idSoliCompra=DID0RfCl_XpGnVtVUqGUD1jfFosOdciHdB_n-LmL_rw (accessed in october 2021).

[31] ELECAUSTRO, *"Informe de evaluación de las ofesrtas presentadas dentro del procedimietno de selección en el exterior para la contratación del Segmento 1 de la*

Fase DOS B del Proyecto Eólico Minas de Huascachaca," [*"Evaluation report of the offers submitted within the selection procedure abroad for the contracting of Segment 1 of Phase TWO B of the Minas de Huascachaca Wind Project,"*] 2020, Available At: https://www.compraspublicas.gob.ec/ProcesoContratacion/compras/PC/informacionProcesoContratacion2.cpe?idSoliCompra=DID0RfCl_XpGnVtVUqGUD1jfFosOdciHdB_n-LmL_rw (accessed in december 2021).

[32] ELECAUSTRO, *"Contrato Nro* [*Contract No.*] *2021 -0023,"* https://www.compraspublicas.gob.ec/ProcesoContratacion/compras/PC/informacionProcesoContratacion2.cpe?idSoliCompra=xFKsUgu1YsQZww5M9IYkBQfTBOpBtWJroZIQa2msIlo, Available At: (accessed in october 2021).

About the Editors

Dr. David Borge-Diez
Department of Electrical, Systems and Automation Engineering.
Electrical Engineering.
University of León, Spain

Daniel Icaza
Researcher
Catholic University of Cuenca-Ecuador

List of Contributors

Mustafa Şahin
İlkay Yavrucuk
Jesús Ordoñez-Guerrero
Semaria Ruiz
Daniel Icaza
David Borge-Diez
Semaria Ruiz
Santiago Pulla-Galindo
Francisco José García-Moya
Enrique Rosales-Asensio

Index

A

annual global electrical performance, 68
annual wind or energy potential production, 67
anti-windup, 56, 104, 106, 115, 116, 117, 123, 127, 129, 135, 136, 137, 138, 140, 141, 142, 143, 144
atmospheric wind, 69

B

batteries, v, 103, 104, 105, 111, 115, 116, 117, 122, 123, 130, 136, 137, 138, 140, 141, 142, 154, 167
battery storage, 105, 106, 109, 111, 129, 150, 177
Blade Element Momentum (BEM) theory, 1, 2, 4, 5, 30
blade pitch controller, 32, 35, 36, 38, 40, 46, 47, 52, 54, 56, 57, 60, 61, 65

C

closed-loop poles, 106, 113, 115
contact voltage drop, 161

D

desalination wind, 182
distribution, v, 67, 79, 98, 100
dynamic and steady-state turbine performance, 35

E

electrical energy hub, 182
energy, 67, 144, 148, 195, 196
energy conversion, 144, 148, 195, 196

F

feedback control, 103, 104, 106, 115, 127, 141, 144
frequency control, v, 103, 104, 105, 106, 107, 108, 109, 110, 111, 118, 119, 123, 127, 129, 135, 141, 142, 143, 144, 145
frequency regulation, 105, 106, 109, 112, 115, 117, 118, 122, 137, 141, 142, 144, 145

G

generated electromotive force, 161
generator torque controller, 15, 35, 36, 39, 40, 43, 46, 47, 48, 50, 51, 52, 56, 57, 61
gross and net production, 68, 101

H

Huascachapa, 203
hub and tip losses, 1, 4
hybrid management system, 182
hybrid plant, 186

I

implementation, v, 1, 3, 4, 15, 16, 30, 66, 102, 111, 116, 144, 159, 171, 175, 179, 199, 201, 205, 206, 208, 210, 213, 215, 217, 222, 223, 224
input voltage to the controller, 162

M
[illegible]
MS Bladed Model, [illegible]

N
nonlinear optimization, 104, 106, 107, [illegible]

O
operation principle, 1, 3, 4, 5, 30, [illegible]
optimization, [illegible]
organic [illegible], 147, 148, [illegible]
organic [illegible], 148, [illegible]
166, 167, 169, 172

P
pitch, 32, 35, 36, 38, 40, 46, 47, 52, 54, 56, 57, 60, 61, 65
probability, 78
probability distribution, 78

R
renewable energy, 16, 30, 31, 32, 64, 65, 66, 109, 112, 113, 144, 147, 148, 150, 151, 152, 153, 154, 155, 156, 157, 166, 173, 174, 175, 176, 177, 178, 181, 182, 184, 185, 186, 187, 188, 189, 190, 191, 192, 193, 194, 196, 201, 203, 209, 210, 216, 217

S
sensor constant (wind vane), 162
skewed wake, 1, 3, 4, 5, 11, 12, 16, 30, 41
state, v, 35, 103, 104, 105, 106, 112, 113, 115, 117, 143
state feedback, v, 103, 104, 105, 106, 112, 113, 115, 117, 143
statistical, 67, 68, 78, 79, 80, 81, 84, 85, 95, 100, 101, 176
statistical analysis, 78, 79, 176
sustainable energy storage, 185, 192

T
[illegible]
turbulent wake, 1, [illegible], 41, 46

U
utility voltage [illegible]

V
validation, 11, [illegible]
velocity, [illegible], 48

W
water [illegible]
Weibull distribution, v, 67, 78, 85, 86, 89, 90, 92, 93, 95, 98, 100
Weibull distribution [illegible], 98, 100
Weibull distribution methodology, v, 67, 98
Weibull parameters, 87, 88, 89, 98
Weibull's Distribution Law, 78, 84
wind, v, vii, 1, 2, 3, 4, 5, 6, 7, 8, 9, 12, 13, 14, 15, 16, 17, 18, 19, 20, 21, 22, 23, 24, 26, 28, 29, 30, 31, 32, 33, 35, 36, 37, 38, 39, 40, 41, 42, 46, 47, 49, 51, 52, 53, 54, 55, 56, 57, 58, 59, 60, 61, 62, 63, 64, 65, 66, 67, 68, 69, 72, 73, 74, 75, 76, 77, 78, 79, 80, 81, 82, 83, 84, 85, 86, 87, 89, 90, 91, 92, 93, 94, 95, 96, 97, 98, 99, 100, 101, 102, 103, 104, 105, 106, 107, 109, 110, 111, 115, 118, 119, 120, 121, 123, 126, 129, 130, 131, 132, 135, [illegible], 142, 143, 144, 145, 147, [illegible], 151, 153, 158, 166, [illegible], 167, 168, 169, 171, 172, 174, 176, 179, 181, [illegible], 187, 188, 189, 190, 191, 192, [illegible], 194, 195, 196, 197, 199, 200, [illegible], 205, 206, 208, 210, 212, 218, 219, 220, 221, 222, [illegible]
wind array, 200
wind data treatment, 80, 98